Designing Out Failure

Designing Out Failure is a comprehensive textbook that introduces a unique, domain-independent methodology for eliminating failure modes and improving reliability across all engineering disciplines.

The book presents a systematic innovative framework for risk reduction and reliability enhancement through a structured methodology applicable to diverse engineering fields. Over 60 worked examples and case studies, spanning mechanical, electrical, structural, and environmental engineering domains, showcase creative solutions of exceptional quality. Each case study is structured not only to demonstrate the principles in action but also to provide actionable insights for one's own design for reliability efforts. Each chapter is complemented by comprehensive practical exercises, tutorial questions, and detailed solutions that reinforce learning and foster creative thinking.

Designed specifically for undergraduate and postgraduate engineering curricula, this textbook teaches creative thinking in engineering design while building core competencies in reliability engineering. It serves as an invaluable resource for students, educators, practicing engineers, and researchers seeking to master the art and science of reliable design.

Designing Out Failure

A Universal Approach to Engineering Reliability

Michael T. Todinov

CRC Press
Taylor & Francis Group
Boca Raton London New York

CRC Press is an imprint of the
Taylor & Francis Group, an **informa** business

Designed cover image: Michael T. Todinov

First edition published 2026
by CRC Press
2385 NW Executive Center Drive, Suite 320, Boca Raton FL 33431

and by CRC Press
4 Park Square, Milton Park, Abingdon, Oxon, OX14 4RN

CRC Press is an imprint of Taylor & Francis Group, LLC

ISBN: 9781041139348 (hbk)
ISBN: 9781041154136 (pbk)
ISBN: 9781003679301 (ebk)

DOI: 10.1201/9781003679301

Typeset in Times
by Newgen Publishing UK

Dedication

To my wonderful wife Prolet

Contents

Author

Michael T. Todinov is a professor of mechanical engineering at Oxford Brookes University, UK, where he teaches reliability engineering, engineering mathematics, and advanced stress analysis. He holds a PhD in mechanical engineering and a higher doctorate, equivalent to a DSc, in mathematical modelling. Prof. Todinov has pioneered innovative research in several areas, including domain-independent methods for reliability improvement, reverse engineering of algebraic inequalities, analysis and optimization of repairable flow networks, reliability analysis based on the cost of failure, and probabilistic fracture controlled by defects. Michael Todinov is a recipient of a prestigious award from the Institution of Mechanical Engineers (UK) for his work on risk reduction in mechanical engineering.

Preface

Reliability improvement is an essential component of product development. Indeed, product development is an iterative process in which the design phase must be followed by reliability verification, and prototype production must be followed by reliability testing.

A solid understanding of failure modes and mechanisms, combined with knowledge of domain-independent principles to prevent them, is fundamental to ensuring the reliability of engineering products. This understanding strengthens design decisions early in the development cycle, preventing costly rework and redesign after the product enters operation.

The most critical mistake a design engineer can make is to overlook reliability during the design stage. The inevitable result is a poorly designed unreliable product, causing customer dissatisfaction, loss of market share, and ultimately, business failure. In today's competitive marketplace, product reliability is not optional, it is vital for survival.

While areas such as reliability and risk assessment are widely acknowledged as domain-independent, the equally critical areas of reliability improvement and risk reduction have not received the same cross-domain recognition. For decades, reliability and risk practitioners have largely failed to recognize that improvements in reliability, reductions in risk, and the management of uncertainty can be guided by general principles applicable across various domains.

With few exceptions, such as the widely recognized methods like introducing redundancy, strengthening and upgrading weak components, selecting better materials, simplifying components and systems, eliminating a common cause, and condition monitoring, there is a lack of a unified framework of domain-independent methods for reliability improvement and risk reduction. This gap has hindered the adoption of broadly applicable strategies that transcend domain boundaries.

Why has engineering design been so slow to embrace the domain-independent methods for reliability improvement and risk reduction?

One of the chief contributing factors is that the prevailing approach remains heavily dependent on domain-specific knowledge and expertise. Many practitioners view domain-independent methods as too generic to tackle the specific challenges unique to their fields. As a result, efforts to improve reliability remain fragmented, restricted to narrow technical areas and heavily dependent on domain experts.

In contrast, foundational sciences such as mathematics and physics, did not remain bound to the domain-specific solutions from which they emerged. Their transformation into powerful scientific disciplines was driven by the ability to generalize by extracting universal principles from diverse examples and forming structured frameworks with strong predictive power.

The persistent belief that reliability improvement must rely solely on domain-specific methods and experts has led to systemic inefficiencies. These include missed opportunities for cost-effective reliability gains, recurring design errors and the continued presence of unreliable products and processes. Consequently, industries often face suboptimal designs and failures that could have been prevented.

This siloed approach to risk reduction has also led to repeated reinvention, as practitioners independently arrive at the same solutions in isolation. In an era of rapid innovation, where domain-specific knowledge can quickly become obsolete, domain-independent methods offer a compelling alternative. These higher-level approaches are adaptable, capable of guiding effective risk reduction even in the absence of detailed failure histories, reliability data or exhaustive analyses.

Domain-independent methods do not rely on historical failure data or detailed knowledge of specific failure mechanisms, which makes them particularly valuable in the design of new products and processes. Often, they enable substantial risk reduction at minimal or no additional cost. A key principle of domain-independent reliability improvement and risk reduction is the recognition of recurring patterns in reliability problems and their solutions across different fields of human activity. To support this cross-domain application, this book proposes a general algorithm for enhancing reliability and reducing risk using universal methods that apply across a wide range of disciplines.

The proposed domain-independent methods are not intended to replace domain-specific approaches, but to complement them. When integrated with specialized knowledge, domain-independent methods lead to more robust and innovative design solutions.

This book introduces a selection of domain-independent methods for improving the reliability of engineering products and processes. Reliability, simply defined, is the absence of failure and the goal of this book is to present a general methodology for preventing product and process failures.

Through numerous case studies and tutorial exercises based on real-world examples, readers will learn to apply domain-independent principles to address practical reliability challenges. This text is intended as a foundational resource on reliability improvement and risk reduction through universal methods.

By following the methodology presented, readers will be able to systematically identify and eliminate failure modes in selected products and processes, thereby significantly enhancing their reliability. These methods have been tested extensively in classroom settings at both undergraduate and postgraduate levels and have consistently demonstrated their effectiveness. After learning the foundational principles, students have gone on to develop innovative reliability solutions, many of which have been considered patentable.

This book does not aim to provide exhaustive coverage of all known domain-independent methods for improving reliability and reducing risk. Instead, the purpose is to structure the material so that it can support an initial introduction to these methods and classroom instruction. This is achieved through a relatively small subset of domain-independent methods by demonstrating how they can be effectively applied to eliminate or mitigate failure modes from products and processes.

This book is designed to benefit academic researchers in engineering design, quality and reliability engineering, risk and safety, and quality management. It is also intended for professionals and researchers across multiple disciplines, including mechanical, civil, electrical, and environmental engineering. The primary audience includes undergraduate students studying engineering design, MSc students in taught programs, and postgraduate students in reliability engineering and risk management.

In conclusion, I would like to express my deep gratitude to Katya Seward, Mechanical Engineering Editor at CRC Press, Taylor & Francis Group, and Suriya Rajasekar, Senior Project Manager at Newgen KnowledgeWorks, for their invaluable assistance and support. I am also thankful to my academic colleagues for their insightful feedback on various aspects of this work.

Finally, I extend my deep gratitude to my wife, Prolet, for her unwavering support, strength, encouragement, and patience during the preparation of this book.

Michael T. Todinov
Oxford, August 2025

1 Basic Concepts

1.1 LIKELIHOOD OF FAILURE, CONSEQUENCES FROM FAILURE, AND RISK OF FAILURE

The purpose of risk analysis is to support informed and effective management decision-making. By evaluating the risks associated with various decision alternatives, risk analysis facilitates the identification of the option that maximizes expected utility for stakeholders, while satisfying specified criteria and constraints.

Core concepts in risk management include the *probability of failure*, *consequences of failure*, and *risk of failure*. Risk is generally associated with adverse events, which are events that have a negative impact.

Although numerous definitions and interpretations of risk exist, for instance, risk as uncertainty about system performance (Aven, 2003), or as a measure of loss variability (Crouhy et al., 2006), most definitions share three fundamental characteristics (Kaplan and Garrick, 1981): (i) A loss-generating factor (an event with negative impact), (ii) The likelihood of exposure to this factor, and (iii) The consequences of the exposure, including the uncertainty surrounding them.

A *hazard* is any situation, condition, or object with the potential to cause harm. Common sources of hazards include systems or materials containing significant amounts of electrical, chemical, thermal, kinetic, or potential energy. Examples include: flammable or highly reactive substances, toxic chemicals or wastes, pressurized gases or explosives, materials at extreme temperatures, high-voltage electrical equipment, electromagnetic fields, high-intensity radiation, loud noise, vibration, fast-moving objects, sharp objects, and suspended heavy objects.

When certain triggering conditions are met, hazards can lead to adverse events.

Losses resulting from adverse events can typically be grouped into several major categories:

- *Loss of life or health impacts*;
- *Environmental and infrastructure damage*;
- *Financial losses*, such as warranty costs, production downtime, asset loss, intervention and repair costs, legal penalties, medical expenses, and insurance claims;

DOI: 10.1201/9781003679301-1

- *Reputational damage*, including loss of customers, contracts, market share, investor confidence, and brand value.

Depending on the type, these losses may be quantified in monetary terms, number of fatalities, downtime, production volume loss, pollution volume, lost customers, or revenue. In many cases, especially in engineering and finance, losses from failures are expressed in monetary terms and referred to as the *cost of failure*.

Different domains exhibit distinct failure modes, each with specific risk characteristics. Fundamentally, risk is defined as a combination of:

- *An adverse event,*
- *The probability of its occurrence, and*
- *The associated loss, conditional on the event occurring.*

Risk should not be equated merely with the magnitude of potential loss, whether fixed or variable. Without exposure to an adverse event, loss cannot materialize.

Unlike alternative neutral viewpoints of risk (Giddens, 1999) where risk could also be associated with potential benefits, the risk in the present treatment is understood to be always associated with potential loss. Risk is always understood as a unity of an adverse event with the potential to cause damage/loss, the likelihood of its occurrence and the magnitude of the consequences given that the event materializes. Failures of equipment, systems, operations, and processes are such adverse events. They are inevitably associated with financial losses, damage to health, fatalities, and damage to the environment.

According to a classical formulation (Henley and Kumamoto, 1981; Vose, 2000), *risk of failure* is quantified as:

$$K = p_f C \tag{1.1}$$

where p_f is the probability of failure and C is the cost given failure.

For an operator of production equipment, the cost associated with a failure, denoted by C, may encompass multiple components. These include the cost of lost production, environmental cleanup, medical expenses, insurance claims, legal liabilities, mobilization of emergency resources, and lost sales resulting from reputational damage and diminished customer confidence.

From the perspective of the manufacturer of the production equipment, the cost of failure may involve warranty payments (if the failure occurs within the warranty period), lost future sales, penalty fees, compensation payments, and legal expenses.

The theoretical basis for Equation (1.1) can be established through the following thought experiment. Suppose that a non-repairable equipment is put in operation for a length of time which can be the warranty period for the equipment. If the equipment fails before the specified warranty time, its failure is associated with a constant loss C due to warranty replacement. Suppose that the experiment involves N identical pieces of equipment, of which N_f fail before the warranty time period. Since failure is associated with losses, the total loss generated by failures among N pieces of equipment is $N_f \times C$. The average (expected) loss K is then $K = (N_f \times C) / N$. This expression can be rearranged as

$$K = (N_f / N) \times C \tag{1.2}$$

If the number of pieces of equipment N is sufficiently large, $p_f \approx N_f / N$ gives an estimate of the probability of failure of the equipment before the warranty period and the expression (1.1) is obtained.

Usually, a relatively small number N gives already a sufficiently accurate estimate of the true probability of failure. As a result, Equation (1.1) represents the average (expected) warranty loss per individual piece of equipment. This helps explain why, in engineering contexts, risk is often equated with the expected loss due to failure. A more detailed discussion of risk, potential loss, the limitations of using expected loss as a measure of potential failure consequences, and common misconceptions about risk can be found in Todinov (2006a; 2006b, 2007).

1.2 FAILURE MODES AND RISK REDUCTION

Reliability is defined as the ability of a system or component to perform required functions without failure under specified conditions, for a given period of time. It is typically measured by the probability of surviving a specified time interval.

Sustainability implies reliability.

- A system that is not reliable cannot be sustainable. It fails prematurely and leads to increased costs for repair or replacement. In the long run, systems that frequently fail are not sustainable from an economic perspective.
- Sustainable practices emphasize the efficient use of resources. Reliable designs align with sustainable principles by minimizing the waste of resources needed to replace failed products.
- Sustainable systems inherently aim to minimize environmental harm (e.g., emissions, water pollution, or damaged infrastructure). A system that fails and causes environmental damage cannot be considered sustainable.
- Systems not designed for reliability, especially under changing conditions, lack resilience to shocks and stresses, and therefore cannot be sustainable.
- Sustainable engineering is not solely about the environment; it also encompasses societal impacts. Systems and processes that are unsafe for users cannot be considered sustainable.
- Systems that lack reliability do not exhibit predictable behaviour, such as consistent performance, returns, or costs, and thus cannot be sustainable in the long run.
- Unreliable systems do not result in long-term cost savings, which are essential for sustainability as resources become scarcer and environmental regulations grow stricter.

- Unreliable products and operations erode the trust of consumers, investors, and partners. This leads to reputational damage, reduced brand loyalty, and ultimately, to a loss of business sustainability.

Failure occurs when the functions delivered by a system or component deviate from the required ones.

A failure mode is the manner in which a system or component fails to perform its intended functions. It represents the observable effect or manifestation of a failure. In other words, it is how a system, component, process, or operation ceases to operate as designed.

Common failure modes of engineering products include: *fracture, excessive wear, yielding, buckling, excessive elastic or plastic deformation, excessive noise or vibration, corrosion, leakage, overheating, lack of precision, stress relaxation, jamming*, etc.

In *project management*, for example, failure modes can also be identified, such as: *unavailability or delay in acquiring key resources* (materials, equipment, software, trained personnel, etc.), *loss of key project personnel, third-party failure to deliver critical project tasks, changes in project scope, delays in receiving training for essential tasks, equipment breakdowns, failure to access databases or software, poor-quality measurements due to faulty equipment*, etc.

The consequences of these failure modes typically include: failure to achieve project objectives, inferior quality, missed deadlines, and excessive project costs.

Most failures in engineering products and processes stem from *variability in critical design parameters*, such as material properties, applied loads, and component strength.

A *failure mechanism* refers to the physical process or sequence of events leading to a particular failure mode. The same failure mode, such as fracture, can arise from different failure mechanisms. For example, *the brittle fracture mechanism* is characterized by little or no plastic deformation. The crack propagates rapidly once initiated, without the need for increased stress. The component fails suddenly, without warning.

The *ductile fracture mechanism* involves significant plastic deformation before failure. The local stresses are redistributed and relaxed through deformation. Ductile fracture requires more strain energy to develop, and the plastic deformation enhances material strength through strain hardening. The crack propagation typically requires an increased load which provides early warning and time for intervention. This makes ductile materials preferable in safety-critical applications.

The *fatigue fracture mechanism*, which occurs under cyclic or varying loads, involves slow crack initiation and propagation over time. The crack growth rate is governed primarily by stress range rather than maximum stress. Once the crack reaches a critical size, final failure occurs by brittle or ductile fracture.

The *creep fracture mechanism* occurs under constant load at elevated temperatures and is primarily caused by grain-boundary sliding. It involves very slow, time-dependent deformation. The rate of deformation depends on the material, temperature, stress level, and duration of exposure. Failure happens when deformation exceeds the critical limit, rendering the component incapable of functioning.

Thus, although the failure mode in each of these cases may be *fracture*, the underlying failure mechanisms are fundamentally different.

Beyond fracture, other failure modes include: *short circuit*, *open circuit*, *overheating* of electrical or mechanical components, *leakage* from seals, *misalignment causing loss of precision*, *excessive deformation*, *contamination*, *excessive noise*, and *vibration*.

An essential part of the risk management process is identifying, preventing or mitigating potential failure modes.

Design for reliability is mostly about preventing failure modes from occurring during the specified lifetime of the product.

The traditional assessment of individual failure modes is done by using *a risk matrix* (Vose, 2000). The likelihood of failure and the consequences given failure are expressed qualitatively, for example on a scale 1 to 5 (1 – very low, 2 – low, 3 – medium, 4 – high, 5 – very high). Next, for each individual failure mode, a risk score is determined as a product of the likelihood ranking and the consequences ranking:

Risk score = Likelihood ranking x Consequences ranking

Ranking failure modes is done according to their risk scores. The risk scores are used to prioritize failure modes. Failure modes with a high score should be given a high priority.

According to the *Pareto principle*, a small number of failure modes is responsible for most of the risk associated with a system or process. For these failure modes then risk reduction measures are applied. The risk management procedure involving failure modes consists of the following steps:

1. Identification of failure modes
2. For each failure mode:
   - estimate qualitatively the likelihood of its occurrence
   - estimate qualitatively its consequences (impact)
   - determine its risk score
3. Rank the failure modes according to their risk scores
4. Select risk reduction/mitigation measures for the failure modes with large risk scores.

The risk reduction measures reduce either the likelihood of failure, the consequences of failure or both (Figure 1.1). Accordingly, they can be preventive measures reducing the likelihood of the failure mode, for example by redesigning the product or process. They can also be protective measures, reducing the severity of the failure mode, for example by introducing deliberate weakness or by fail-safe design. Finally, they can be dual measures, reducing both the likelihood of failure and the severity of the consequences (Todinov, 2007).

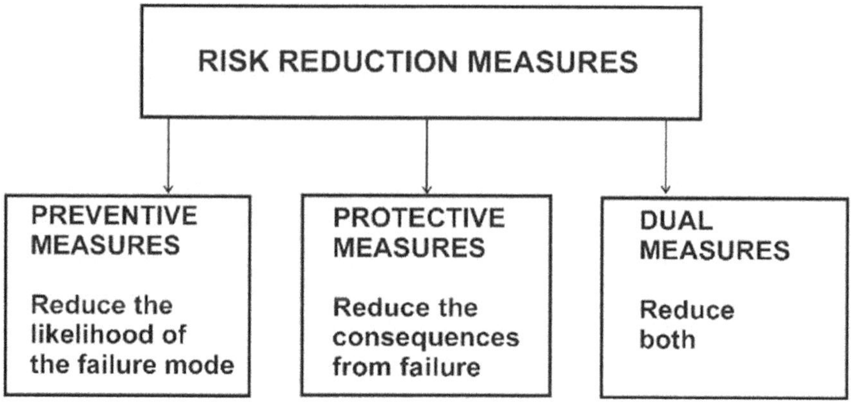

FIGURE 1.1 Risk reduction measures.

1.3 TECHNIQUES FOR IDENTIFYING AND ASSESSING FAILURE MODES

The risk of failure is primarily reduced by eliminating potential failure modes. Design for reliability focuses on preventing failure modes from occurring during the specified lifetime of the product. This involves systematically identifying as many potential failure modes as possible and assessing their impact on overall system performance. The goal is to highlight critical areas where design modifications can either lower the likelihood of failure or mitigate the failure's consequences. By doing so, limited resources for risk reduction can be directed effectively toward the most vulnerable aspects of the design.

It is essential to identify failure modes and implement the necessary design modifications early in the development process, before these issues manifest in the field. Changes made during the early design stages are significantly more cost-effective than those implemented later. *Although early detection of failure modes can be challenging, addressing them at an early stage is relatively straightforward. Conversely, once the product is released, failures are easier to detect but far more difficult and costly to eliminate.*

Therefore, to minimize the risk of failure in a product or process, it is crucial to identify potential failure modes as early as possible. This enables timely implementation of design changes and targeted actions that reduce risk.

Structured methodologies have been developed to facilitate the identification and evaluation of failure modes. One of the most widely recognized is Failure Mode and Effects Analysis (FMEA), introduced by NASA in 1963 for the Apollo program. FMEA has since been adopted across various industries, including aerospace, nuclear, electronics, automotive, and software development. Much of the foundational literature on FMEA originates from the American Military Standard MIL-STD-1629A (1977).

FMEA is designed to uncover as many potential failure modes as possible during the early design phase. It evaluates their consequences, investigates root causes, and recommends controls and corrective actions to minimize associated risks. When

FMEA is expanded to include a criticality assessment, it becomes Failure Mode, Effects, and Criticality Analysis (FMECA).

FMEA employs an inductive, bottom-up approach to risk analysis (McDermott et al, 2008). It systematically examines the failure modes of each component or subsystem and traces their effects on overall system performance. This sequential analysis ensures a comprehensive understanding of potential weaknesses, enabling more robust and resilient designs.

1.3.1 Case Study: Discovering the Failure Modes of a Lubrication System by Using an Inductive (Bottom-Up) Approach

Before discovering any failure modes is attempted, it is vital to understand the basic processes in the lubrication system and how the system works. In this respect, building a functional block diagram and specifying the required functions of the lubrication system is very important. The functional diagram shows how the components or process steps are interrelated (Todinov, 2016).

For example, the required function from the generic lubrication system in Figure 1.2 is to supply constantly clean oil at a specified pressure, temperature, debit, composition, and viscosity to contacting moving parts in order to: (i) reduce wear; (ii) remove heat from friction zones and cool the contact surfaces; (iii) clean the contact surfaces from abrasion particles and dirt; (iv) protect from corrosion the lubricated parts.

These functions are guaranteed by using components with specific functions. The sump is used for the storage of the oil. The oil filter and the strainer are used

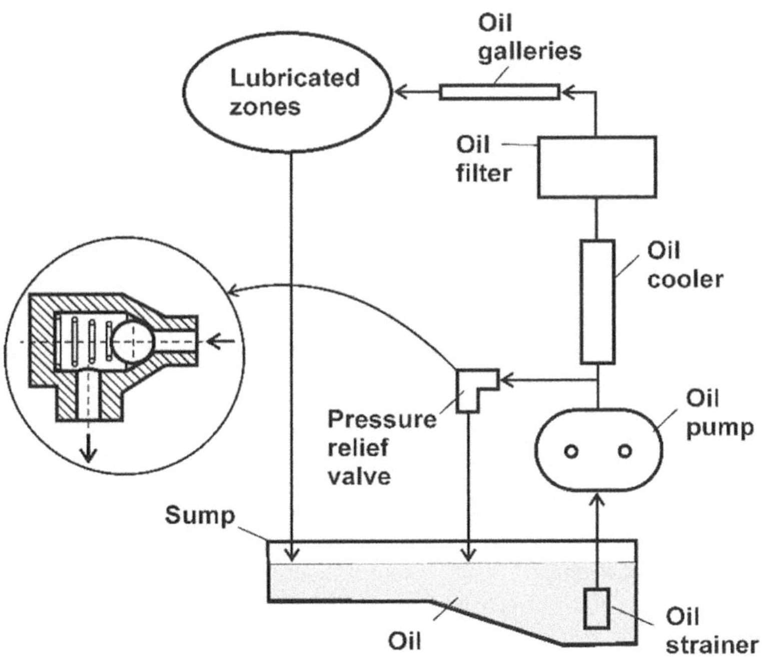

FIGURE 1.2 Functional block diagram of a lubrication system.

to maintain the oil cleanliness. The pressure relieve valve is used for maintaining the oil pressure. The oil cooler is used for maintaining the oil temperature. The oil pump is used for maintaining the oil debit and the oil galleries are used for feeding the oil where it is needed. Negating these functions reveals important failure modes.

The inductive approach for discovering failure modes starts from the failure modes of the separate components and tracks their impact on the system's performance.

Thus, a clogged filter in the system from Figure 1.2 leads to a significant pressure difference before and after the filter and low pressure of the lubricating oil. A mechanical damage of the oil filter prevents the retention of suspended particles in the oil and leads to a loss of the function 'supply of clean oil to the lubricated surfaces'.

If the pressure relief valve is stuck in open position, the oil pressure cannot build up and the pressure after the oil filter will be low. This will lead to a low pressure in the supplied oil. If the valve is stuck in closed position, this will lead to the system failure mode 'too high oil pressure'. The oil pressure will steadily build up, and with no pressure relief mechanism, the high pressure could destroy the oil filter and even blow out the oil plugs.

An oil cooler lined up with deposited plaques or clogged with debris is characterized by a reduced heat transfer coefficient which leads to decreased cooling capability and a 'high temperature of the oil in the system'. Failure of the cooling circuit will have a similar effect. Clogging the cooler with debris will simultaneously lead to an increased oil temperature due to a decreased cooling capability and a pressure drop across the cooler which results in a low oil pressure in the lubrication system.

Excessive wear of the oil pump leads to low oil pressure while a broken oil pump leads to no oil pressure. Failure of the sump leads to no oil pressure; blocked oil strainer leads to a low oil pressure. Blockage of the oil galleries, badly designed oil galleries or manufacturing defects lead to loss of the function 'delivering oil in the specified debit where it is needed'.

Oil contamination due to inappropriate storage, oil degradation caused by oxidation or depletion of additives, and the selection of inappropriate oil, lead to a failure to supply oil according to the specified composition and viscosity.

In summary, the inductive analysis reveals the following failure modes at a system level:(i) no oil pressure; (ii) too low oil pressure; (iii) too high oil pressure; (iv) too high oil temperature; (v) oil not filtered; (vi) oil contamination; (vii) oil degradation;(viii) not delivering the oil where it is needed; (ix) insufficient amount of oil delivered to lubrication zones.

1.3.2 Limitations of Traditional FMEA

While FMEA is widely used, it has significant limitations. Here are some of these:

- *Incomplete Coverage*
  FMEA offers no assurance that all possible failure modes have been identified.

- *Component Independence Assumption*
  FMEA typically treats failure modes of components independently. This assumption obscures complex, system-level failure modes that only emerge from interactions among multiple component failures.
- *Inability to Capture Parameter Stack-Up Effects*
  FMEA is not equipped to identify failure modes caused by the combined effect of multiple small deviations in reliability-critical parameters.

For example, in a press-fit assembly, elevated temperatures combined with a low coefficient of friction and minimal interference at room temperature can lead to a loss of torque transmission. Such cases are better addressed through virtual testing using Monte Carlo simulations (Todinov, 2016).

- *Inability to Capture Time-Dependent Failures*
  FMEA does not account for failure modes arising from timing or clustering of events.
  For instance, a failure mode like 'insufficient resource supply' (e.g., steam) can occur if multiple production units demand the resource simultaneously, exceeding the generator's capacity. This failure mode depends solely on temporal clustering of demand, a factor not addressed by standard FMEA.

1.4 TOP-DOWN APPROACH

The top-down approach is a basic technique which helps to identify contributing root causes to failure modes. It starts from a top event (effect, system failure mode) and goes down to the root causes of this event. It also considers an already identified failure mode at a system level and investigates what else could cause this failure mode.

1.4.1 Ishikawa Diagrams

This approach can be illustrated with Ishikawa diagrams. The Ishikawa diagram of the failure modes 'no oil pressure', 'too low oil pressure', 'too high oil pressure', 'too high oil temperature', 'oil contamination' and 'oil degradation' are shown in Figures 1.3–1.8.

1.4.2 Fault Trees

Fault trees can also be used to present the root causes leading to a failure mode. These can be built with AND and OR gates (Figure 1.9). Because of the AND and OR gates, the fault trees permit including:

- Causes which are present only if a number of conditions are simultaneously present. This case is described by using an AND gate.
- Causes which are present only if at least one condition (from a set of conditions) is present. This case is described by using an OR gate.

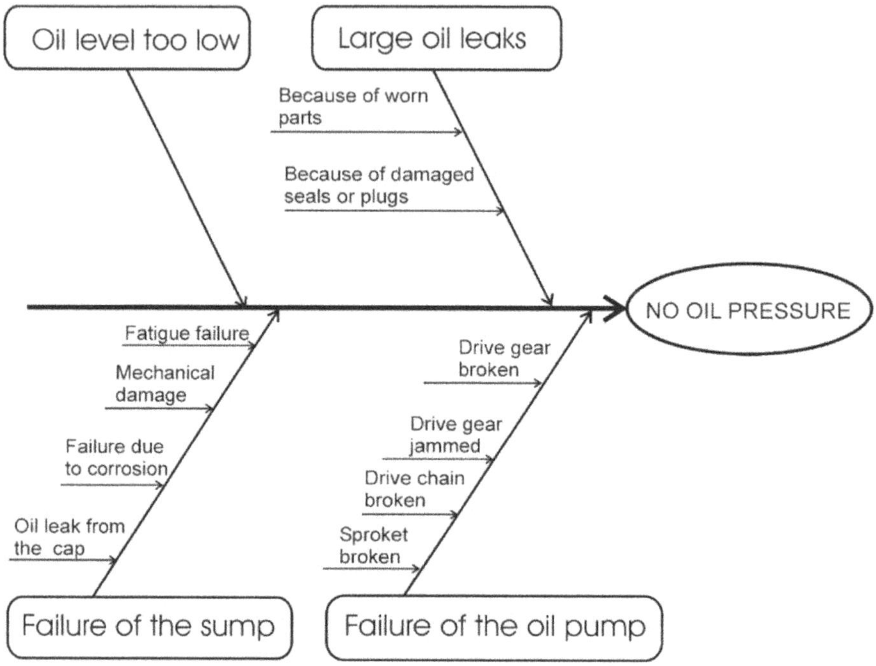

FIGURE 1.3 Ishikawa diagram of the failure mode 'no-oil pressure'.

1.4.3 Case Study: Premature Fatigue Failure of a Compression Spring

The following root causes have been identified for the premature failure of a hot-coiled silicon-manganese automotive suspension spring (Figure 1.10).

1. Design error;
2. Low quality of the spring steel;
3. Excessive decarburization of the spring wire;
4. Inappropriate heat treatment of the coiled spring;
5. Inappropriate shot peening of the spring coil;
6. Corrosion.

The presence of any of these leads to a premature fatigue failure of the suspension spring.

For example, the low quality of the steel can be due to anisotropy of the spring wire, large sulphide inclusions or large oxide inclusions. The presence of any of these leads to a low quality of the spring wire.

The excessive decarburization during heating of the spring wire before coiling is caused by low carbon potential of the atmosphere during heating and insufficient alloying to prevent decarburisation. Only the simultaneous presence of both factors leads to excessive decarburization of the spring wire.

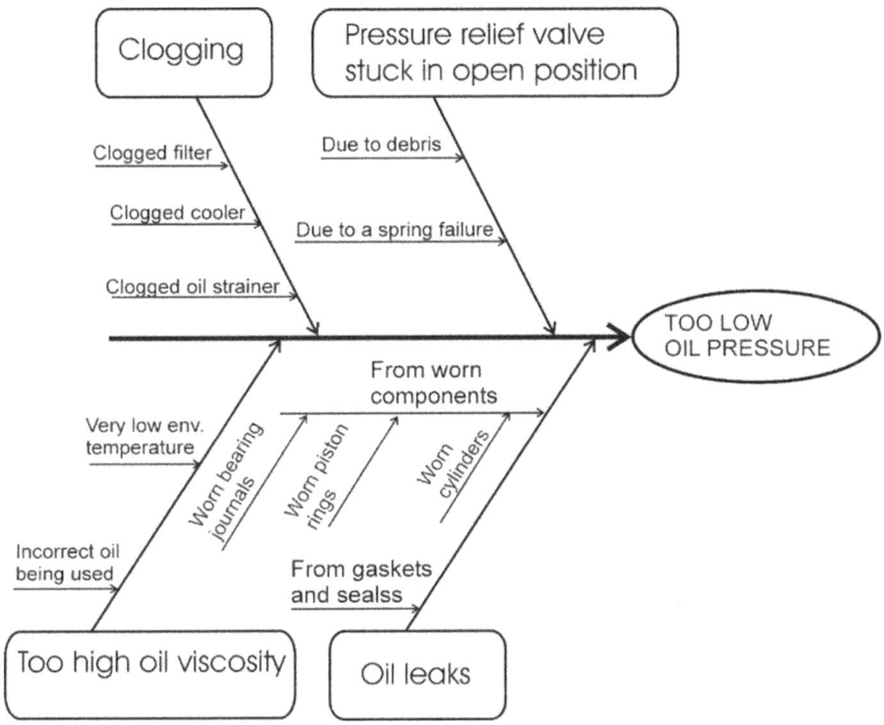

FIGURE 1.4 Ishikawa diagram of the failure mode 'too low oil pressure'.

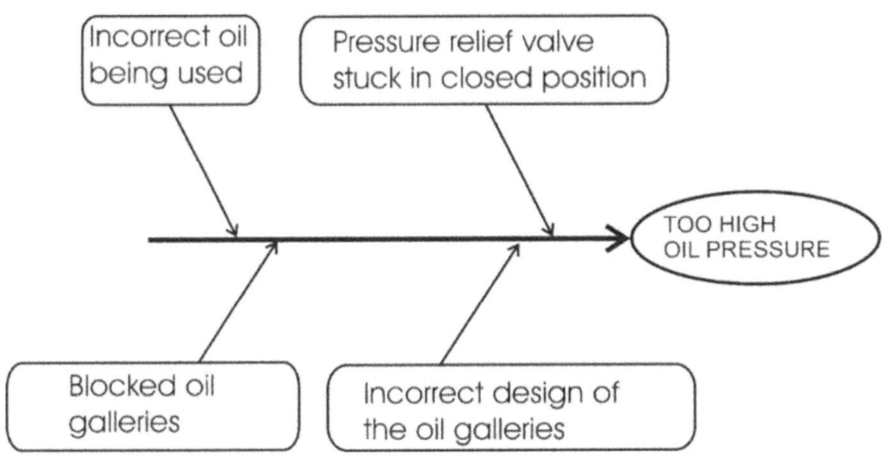

FIGURE 1.5 Ishikawa diagram of the failure mode 'too high oil pressure'.

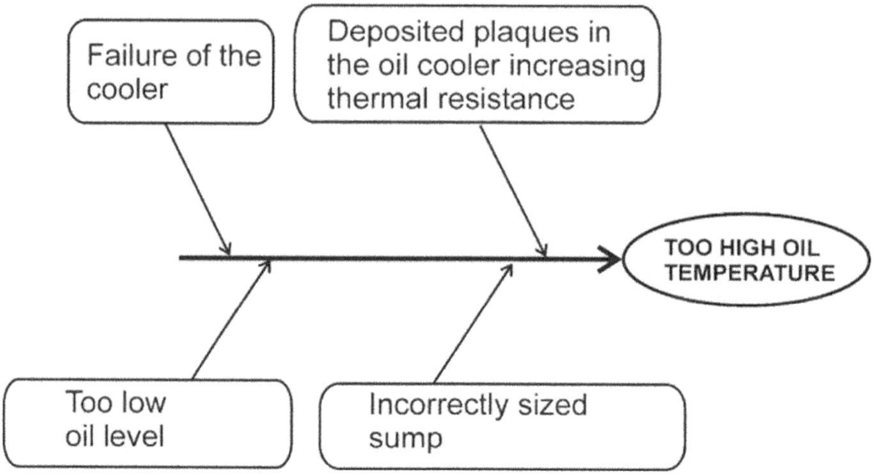

FIGURE 1.6 Ishikawa diagram of the failure mode 'too high oil temperature'.

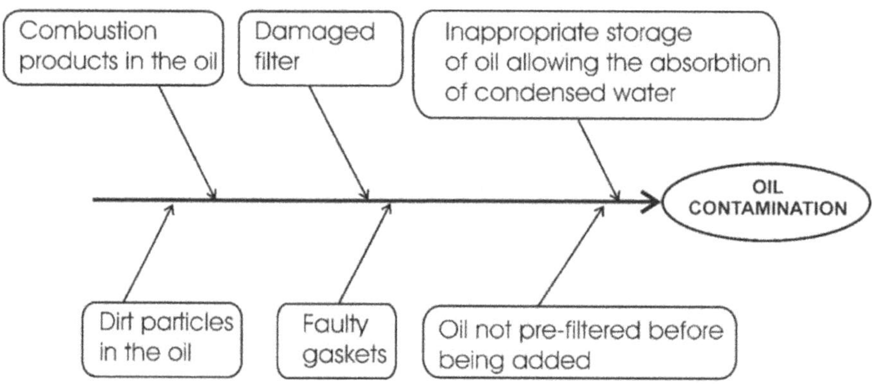

FIGURE 1.7 Ishikawa diagram of the failure mode 'oil contamination'.

The inappropriate shot peening can be due to insufficient intensity of the shot peening unable to ensure a compressive residual stress of large magnitude at the surface or to a non-uniform shot peening leaving zones of reduced residual stress along the circumference of the spring wire.

The inappropriate heat treatment can be due to coarse grain after austenitization, sub-optimal tempering resulting in too soft or too hard spring wire or to large residual tensile stresses after tempering that could not be masked by the compressive residual stresses from shot peening.

An inappropriate heat treatment could also be due to presence of microcracks after quenching which serve as fatigue crack initiation sites.

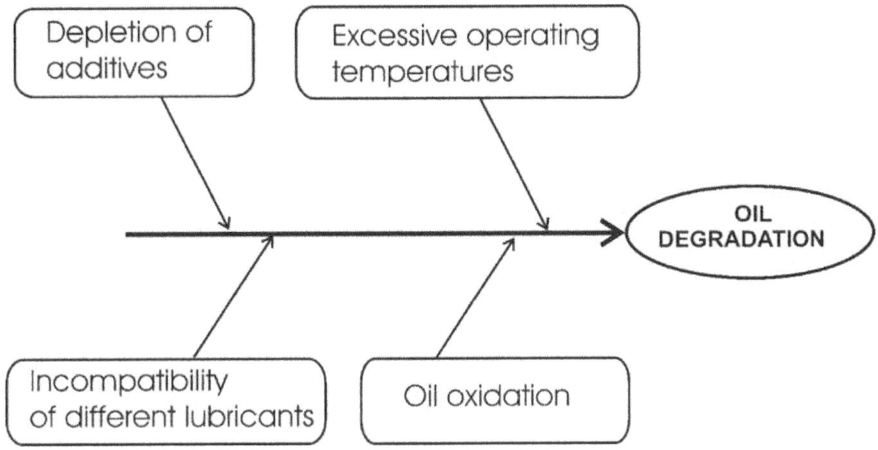

FIGURE 1.8 Ishikawa diagram of the failure mode 'oil degradation'.

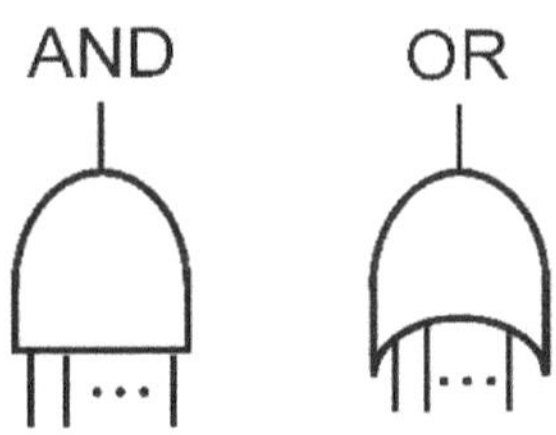

FIGURE 1.9 'AND' and 'OR' gate.

1.4.4 Limitations of the Top-Down Approach

The top-down approach may overlook failure modes at component or subsystem levels that do not directly lead to the top-level event. It also ignores latent or minor failures unless they directly contribute to the top event.

The top-down approach depends on known failures and requires a clear definition of the top-level failure event. It is not very effective if the system failure modes are not well understood in advance. This is why, it has limited discovery of unexpected new failure modes.

It also becomes increasingly complex with system size and interaction. Fault trees, for example, can become cumbersome, deep, and difficult to manage.

The top-down approach also requires high-level expertise and needs expert knowledge to accurately construct the fault trees and understand system interactions. It also assumes linearity and independence and may not accurately capture non-linear interactions or dependent failures between components. Often, the top-down approach does not handle common-cause failures well.

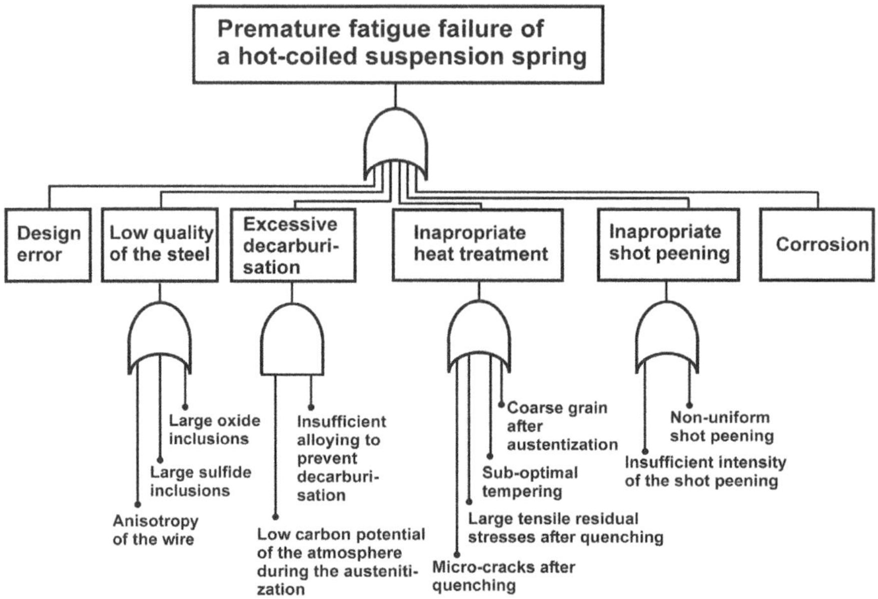

FIGURE 1.10 A fault tree for premature failure of a hot-coiled automotive suspension spring.

1.5 OTHER TECHNIQUES FOR IDENTIFYING POTENTIAL FAILURE MODES

In addition to traditional methods, a variety of techniques can be used to identify potential failure modes in systems and components. Here are some of them.

Systematic Analysis Using Checklists

Structured checklists are useful tools for uncovering common failure modes. A sample checklist for mechanical assemblies includes the following questions:

- Are components sensitive to variations in load?
- Are they resistant to temperature fluctuations?
- Can they withstand vibration?
- Are they resistant to corrosion?
- Are assemblies robust against tolerance variations?
- Are parts sensitive to precise alignment?
- Are components prone to misassembly?
- Are they vulnerable to contamination?
- Are they susceptible to stress relaxation?

Learning from Past Failures
Databases of 'lessons learned' from previous failures are invaluable, particularly in high-stakes industries such as the oil and gas industry, aerospace, and nuclear power. These repositories help prevent failure recurrence by capturing insights from real-world incidents.

Expert Interviews
Engaging with domain specialists through interviews can surface hidden or domain-specific failure risks that may not be evident through analysis alone.

Devil's Advocacy and 'What If' Analysis
Purposefully challenging the design by asking 'what could go wrong?' helps identify edge cases and failure scenarios that might otherwise be overlooked.

Root Cause Analysis
This method uncovers both expected and unexpected failure modes by revealing underlying processes and conditions that lead to failures. It is especially effective in tracing the genesis of complex failures.

Assumption Analysis
Testing the validity of assumptions about design procedures, manufacturing methods, usage, working conditions, and environmental factors often reveals overlooked vulnerabilities.

Constraint Analysis
Examining the technical, operational, and environmental constraints of a system can expose failure modes linked to boundary conditions or system limitations.

Reverse Thinking about Failure
Instead of merely asking what could go wrong, this technique challenges analysts to think about how to intentionally make the system fail. This shift in perspective often exposes rare or unexpected failure modes.

Creative Problem-Solving Techniques
Brainstorming, TRIZ (Theory of Inventive Problem Solving), and lateral thinking are effective tools for uncovering failure modes in products and processes.

For example, Anticipatory Failure Determination (Kaplan et al., 1999) deliberately seeks to precipitate failure using existing system characteristics and environmental factors. This often reveals failure modes that conventional methods miss.

Before attempting to identify failure modes, it is crucial to understand the system's core functions and internal processes. Developing a functional block diagram and clearly defining system functions are foundational steps. These diagrams illustrate interdependencies between components and process steps, enabling a more complete failure analysis.

2 Domain-Independent versus Other Approaches for Reliability Improvement and Risk Reduction

2.1 THE ABSENCE OF GENERAL RELIABILITY IMPROVEMENT METHODS

While reliability and risk assessment are widely recognized as domain-independent disciplines, the same cannot be said for the equally important areas of reliability improvement and risk reduction. For decades, reliability and risk science has failed to acknowledge that improvements in reliability, reductions in risk, and the management of uncertainty can also be guided by general principles that apply across diverse fields. Most progress in reliability improvement and risk reduction takes place within isolated domains, where principles, methods, and techniques are rarely shared. This lack of cross-domain exchange hinders the dissemination of powerful approaches that could accelerate research and development, prevent missed opportunities for enhancing reliability, and reduce the effort wasted on reinventing existing solutions. With few exceptions, such as the widely known domain-independent techniques like introducing redundancy, strengthening weak links, introducing deliberate weak links, upgrading components, simplifying systems and operations, and condition monitoring, the field lacks a comprehensive framework of general methods for risk reduction that transcend specific domains.

Consequently, research and development have largely focused on domain-independent tools for measuring and assessing reliability, risk, and uncertainty, rather than on developing universal strategies for improving reliability or reducing risk and uncertainty that can directly inform the design process. Standard mechanical engineering and design textbooks (e.g., Collins, 2003; Norton, 2006; Pahl et al., 2007; Childs, 2014; Budynas and Nisbett, 2015; Mott et al., 2018; Gullo and Dixon, 2018; French, 1999; Samuel and Weir, 1999; Thompson, 1999; Ugural, 2022) do not mention generic, domain-independent methods for reliability improvement or risk reduction.

Traditionally, reliability improvements and risk reduction have relied on feedback from reliability testing or customer experience. Once a failure mode is

 DOI: 10.1201/9781003679301-2

identified, components are redesigned to address it. However, this approach has major limitations: the feedback typically arrives too late, after the product is manufactured, which makes corrective changes costly or unfeasible. Moreover, reliability testing programs are expensive and significantly increase product cost. In some cases, such as environmental pollution from disposable plastics, the delay in addressing failure modes can have catastrophic consequences.

The ALARP (As Low As Reasonably Practicable) approach to risk management (Cullen, 1990; HSE, 1992; Melchers, 2001) emphasizes reducing risks until the cost of further reduction outweighs the benefits. While this framework supports decision-making through cost-benefit analysis, it offers little guidance on the methods that can be used to reduce risk. For example, Thompson (1999) emphasized integrating maintainability and reliability in design and highlighted the role of failure mode and effects analysis (FMEA). However, while FMEA is valuable for tracing how component malfunctions propagate through a system, it provides no insight into the underlying principles of designing out failure modes.

Thompson (1999) rightly noted that understanding the principles of risk is crucial for achieving reliability, but no set of domain-independent principles for reliability improvement and risk reduction has been formally established. French (1999) proposed several general design principles, but they were not explicitly aimed at enhancing reliability or managing risk. Similarly, Pahl et al. (2007) presented general design guidelines, though many were too specific and lacked broad applicability. Collins (2003) discussed design from a failure prevention standpoint, but without formulating general, cross-domain principles for risk reduction.

The broader literature on reliability and risk (e.g., Bazovsky, 1961; Barlow and Proschan, 1965, 1975; Ang and Tang, 1975; Dhillon and Singh, 1981; Billinton and Allan, 1992; Ramakumar, 1993; Lewis, 1996; Ebeling, 1997; Meeker and Escobar, 1998; Vose, 2000; Booker et al., 2001; Kuo et al., 2001; O'Connor, 2002; Trivedi, 2002; Aven, 2003) has primarily focused on risk modelling, assessment, and decision-making, with minimal attention to domain-independent strategies for reducing technical risk.

A major reason for the neglect of domain-independent principles for reliability improvement is the entrenched belief among domain specialists that their field-specific expertise is sufficient to resolve all reliability challenges. Many domain experts view domain-independent methods as too generic to address the nuanced issues within their specific domains. Consequently, reliability improvement efforts remain confined within narrow domains and are carried out exclusively by domain experts.

This mindset has led to systemic inefficiencies: missed opportunities for effective, low-cost reliability improvements; frequent repetition of known design errors; and the existence of products and processes associated with high risk of failure. As a result, the industry continues to suffer from inferior designs and preventable failures. Petroski (1994) documents numerous catastrophic failures caused by repeated design mistakes, many of which could have been avoided through the application of general, cross-domain principles.

2.2 THE DATA-DRIVEN APPROACH TO RELIABILITY IMPROVEMENT

The *data-driven approach* to reliability improvement leverages empirical data, collected from sensors, user feedback, operational logs, testing, and failure records, to guide decisions aimed at enhancing the reliability of systems, components, and processes. This approach complements traditional physics-based and expert-driven approaches by offering real-world, quantitative insights.

A key advantage of the data-driven approach is its ability to ground decisions in observed data rather than assumptions or subjective judgments. This allows for the identification of the most frequent or severe failure modes, enabling targeted and effective reliability enhancements.

Moreover, data-driven methods facilitate the *early detection of failure patterns*. Techniques such as machine learning and statistical data analysis can be employed to *predict failures before they occur*, uncovering latent degradation trends or fault progressions that might escape traditional inspections. Usage of failure data also creates feedback loops that inform better design decisions.

However, there are important limitations to the data-driven approach. Many established methods for measuring and assessing reliability and risk cannot be readily applied during the design phase, primarily because reliability data is often unavailable for new products or processes.

Additionally, *environmental variability* often undermines prediction accuracy. Models based on data collected under specific environmental conditions (e.g., temperature, humidity, pressure, vibration, corrosive agents) often provide poor reliability estimates when applied to systems operating in different environments.

Even when component-level reliability data is available, recent research (Todinov, 2024b) based on reverse engineering of algebraic inequalities has revealed a fundamental flaw in the current system reliability prediction practices. Due to inherent variability among components of the same type, using *average component reliabilities* to estimate the reliability of series or series-parallel systems is fundamentally flawed, even when components operate independently. Similarly, applying *average hazard rates* to predict the reliability of systems with parallel arrangement leads to incorrect results.

That said, average component reliabilities and hazard rates can still be *useful in comparative analyses*, where two or more system configurations are evaluated against each other in terms of reliability or availability (Todinov, 2009). But for *absolute reliability predictions*, using average values for the reliability of components of the same variety, does not yield correct reliability predictions.

This conclusion, reached through analytical reasoning in Todinov (2024b), had already been supported by the results observed by many reliability practitioners, many of whom became convinced of the futility of absolute system reliability predictions and subsequently abandoned the use of average component reliabilities for system reliability forecasts.

2.3 THE PHYSICS-OF-FAILURE APPROACH TO RELIABILITY IMPROVEMENT

The *physics-of-failure* approach to reliability improvement is grounded in understanding the fundamental mechanisms that cause components and systems to degrade and fail. Developed in response to the limitations of data-driven methods, the physics-of-failure approach (Pecht et al., 1990; Dasgupta and Pecht, 1991; Pecht, 1996) emphasizes reliability prediction based on principles from materials science, mechanics, thermodynamics, and other physical sciences. It enables engineers to model and simulate how products will behave under expected loads, temperatures, and environmental conditions before they are built. This, in turn, supports the estimation of remaining useful life based on stress levels, usage profiles, and environmental exposure.

According to the physics-of-failure approach, failures and performance degradation result from identifiable and quantifiable failure mechanisms. Damage accumulates over time, and failure occurs once this damage surpasses a critical threshold or endurance limit. Consequently, time-to-failure can often be physically modelled with reasonable accuracy. The physics-of-failure approach has proven effective in addressing root causes of failure, eliminating failure modes, and has led many reliability practitioners to view it as the most robust path to genuine reliability improvement.

However, it is equally important to recognize the limitations of this approach. Constructing accurate physics-of-failure models is not always feasible due to the complexity of physical mechanisms, environmental conditions, and operational stresses. Moreover, the physics-of-failure modelling demands extensive and time-consuming research, specialized equipment, and skilled personnel. For example, enhancing component strength through analysis of the relationship between material microstructure and mechanical properties requires significant experimental effort.

More critically, the physics-of-failure models are domain-specific and do not capture the broader principles and invariants that underlie reliability improvement and risk reduction across unrelated fields. Despite their success within targeted domains, the physics-of-failure models cannot be readily generalized and thus cannot drive cross-domain reliability improvement.

2.4 THE TRIZ PROBLEM-SOLVING FRAMEWORK AND THE RELIABILITY IMPROVEMENT

The *TRIZ* (Theory of Inventive Problem Solving) *problem-solving framework* (Altshuller, 1984, 1996, 1999; Salamatov, 1999; Savransky, 2000; Orloff, 2006, 2017; Rantanen and Domb, 2008; Terninko et al., 1998), widely adopted by companies and researchers around the world, has clearly demonstrated the advantages of using *generic principles* to resolve technical contradictions and drive innovation.

- By employing *domain-independent methods*, TRIZ enables rapid mental mapping of complex problems, thereby strengthening intuition and often leading to surprising breakthroughs and accelerated outcomes.

- Reliability problems often involve technical or physical contradictions (e.g., strong but lightweight or simultaneously hot and cold). TRIZ offers tools (e.g., Contradiction matrix, 40 Inventive Principles, ARIZ) to resolve contradictions without compromise and enhance performance without sacrificing durability.
- TRIZ is based on the analysis of a very large number of patents, offering access to a global innovation database and helps engineers apply generalized solutions that have worked across industries.
- TRIZ encourages cross-disciplinary thinking. It can be applied across mechanical, electrical, software, and systems engineering, making it versatile for multidisciplinary reliability challenges. As a result, TRIZ often leads to non-obvious solutions by encouraging the use of principles from other fields, supporting robust, innovative, and fail-resistant product designs.
- TRIZ encourages seeking of *ideal final result*. Its focus on ideal final result pushes designs toward reliability without added complexity or cost, and promotes simplification and robustness.

However, it is important to distinguish between the *generic methods for improving reliability* and the TRIZ methodology. TRIZ is a general framework for inventive problem-solving, not specifically designed for reliability improvement and risk reduction, although some of its 40 inventive principles can, in certain cases, be used for improving reliability and reducing risk. The widespread successful application of TRIZ in industry provides strong evidence of the power of generic thinking in problem solving. Nevertheless, TRIZ *cannot substitute* for the generic methods specifically developed for reliability improvement, for several key reasons:

- *Lack of mathematical modelling framework*: TRIZ does not employ mathematical models or algorithms. In contrast, many reliability problems require formal modelling to devise an effective risk-reduction strategy. Reliability improvement in such cases depends on constructing and comparing reliability models of competing systems to determine which is intrinsically more reliable.
- TRIZ does not provide dedicated tools or methods for handling uncertainty. In contrast, many risk-reduction activities require making decisions with limited or no information about the parameters that govern system or process behaviour. Effective risk reduction in such situations depends on the ability to make sound decisions despite the absence of reliable data about the key controlling variables.
- *Incomplete coverage of reliability principles*: The 40 TRIZ inventive principles do not encompass core techniques that work very well in reliability improvement. For example, TRIZ lacks principles related to *improving system balance*, slowing the *rate of damage accumulation*, and using *deliberate weaknesses* to reduce risk.
- *Lack of theoretical reliability methods*: Techniques grounded in reliability theory, such as *removal or blocking of common-cause failures*, *k-out-of-n redundancy,* and *standby redundancy*, fall outside the scope of the TRIZ methodology.

- Although TRIZ offers a set of inventive principles, the lack of a reliability improvement-oriented classification of various techniques within these principles makes their practical application for reliability improvement challenging. Without such a classification, practitioners must rely heavily on imagination, making it difficult to effectively apply the principles.
- *Limited focus on failure mechanisms*: TRIZ does not typically address the physical mechanisms through which damage develops, nor the techniques used to delay or block these mechanisms.
- A major disadvantage of the TRIZ methodology is also its complexity and steep learning curve. While TRIZ is powerful, its effectiveness is limited by the user's ability to master and apply its sophisticated tools. TRIZ requires deep understanding of 40 inventive principles, contradiction matrices, ARIZ (Algorithm for Inventive Problem Solving), and other TRIZ tools. This methodology is not intuitive for beginners and is time-consuming to apply properly, especially in fast-paced environments. It requires extensive training or facilitation to be used effectively.

2.5 DOMAIN-INDEPENDENT APPROACH TO RELIABILITY IMPROVEMENT AND RISK REDUCTION

Technological advancement is progressing so rapidly that domain-specific knowledge related to reliability improvement and risk reduction often becomes obsolete shortly after it is developed. In contrast, the *domain-independent methods* are higher-order approaches that remain relevant across various fields and can be applied effectively in new, evolving situations.

The systematic distillation, formulation, and classification of domain-independent principles and methods for enhancing reliability and reducing technical risk began in Todinov (2007) and continued in later works (Todinov 2015, 2017a, 2017b, 2019a, 2019b, 2019c, 2024a, 2024b, 2025a, 2025b, 2025d). In Todinov (2007), the domain-independent principles for risk reduction were broadly categorized into three types: *preventive* – aimed at reducing the *likelihood* of failure, *protective* – focused on mitigating the *consequences* of failure, and *dual* – addressing both the likelihood and consequences of failure.

These principles and methods transcend their origin in mechanical engineering and are applicable across a wide range of domains. Examples include methods such as *separation*, *stochastic separation*, *segmentation*, *self-reinforcement*, *inversion*, *reducing the rate of accumulation of damage*, *introducing deliberate weaknesses*, *asymmetrical arrangement of interchangeable redundancies*, and *limiting space and time exposure*. These techniques can be effectively utilized in various unrelated fields.

A core theme in the domain-independent framework for reliability improvement and risk reduction, as elaborated in Todinov (2019a), is the idea that risk reduction is governed by *universal principles*. When these principles are combined with domain-specific knowledge, they can yield highly effective risk-reducing solutions.

Domain-independent methods *do not* depend on historical failure data or detailed knowledge of failure mechanisms. This makes them particularly well-suited for designing new systems where such information is unavailable. In many instances, these methods offer significant risk reduction at no cost or minimal additional cost.

Over the past decade, the domain-independent approach has proven highly effective in educating mechanical engineering students at Oxford Brookes University (UK), at both undergraduate and postgraduate levels. This book does not aim to provide exhaustive coverage of all known domain-independent methods for reliability improvement and risk reduction. Instead, the purpose is to structure the material related to several selected domain-independent methods so that it can be used to demonstrate the power of the domain-independent approach, and support classroom instruction and in-company training for reliability engineers. This is achieved through a relatively small subset of domain-independent methods by demonstrating how these methods can be effectively applied to real-world reliability problems.

The establishment of universally accepted theoretical principles for risk assessment hinges on the development of a common definition of risk that is valid across diverse domains. Likewise, establishing solid theoretical foundations for risk reduction requires the articulation of common domain-independent principles applicable to unrelated areas of human activity. Elevating reliability and risk research to a mainstream scientific discipline demands such foundational principles, especially for the two core components of risk management: *risk assessment* and *risk reduction*.

Domain-independent principles and methods for risk reduction:

- Enhance decision-making in reliability improvement and risk mitigation.
- Contribute directly to the design process by improving product reliability rather than merely evaluating it.
- Enable effective risk reduction across diverse domains, preventing missed opportunities for risk mitigation.
- Provide deeper insights into the existing body of risk reduction techniques.

Domain-independent methods are not a substitute for domain-specific approaches. Rather, when integrated with specific domain knowledge, they contribute to superior design solutions. This book demonstrates, for example, that combining insights from the mechanics of materials with domain-independent strategies such as *segmentation* and *aggregation* can lead to significant improvements in reliability and safety.

2.6 GENERAL APPROACH TO BE FOLLOWED IN RELIABILITY IMPROVEMENT AND RISK REDUCTION BY DOMAIN-INDEPENDENT METHODS

Improving the reliability of an engineering product or process primarily involves eliminating failure modes. A successful reliability improvement solution must preserve the original functions of the product or process while removing its critical failure modes. Furthermore, it should not introduce new critical failure modes or increase complexity and cost.

One of the key principles of the domain-independent approach to reliability improvement is the *recognition of commonality* among reliability problems and their solutions across diverse fields of human activity. For example, a common cause, such as an incorrect maintenance procedure, frequently arises in mechanical engineering, electrical engineering, manufacturing, construction, aerospace engineering, automotive engineering, medicine, and environmental sciences.

Another foundational principle is that while reliability problems may be numerous and varied, the general principles underlying effective solutions are relatively few and broadly applicable. This has led to the development of a set of domain-independent methods for reliability improvement and risk reduction (Figure 2.1).

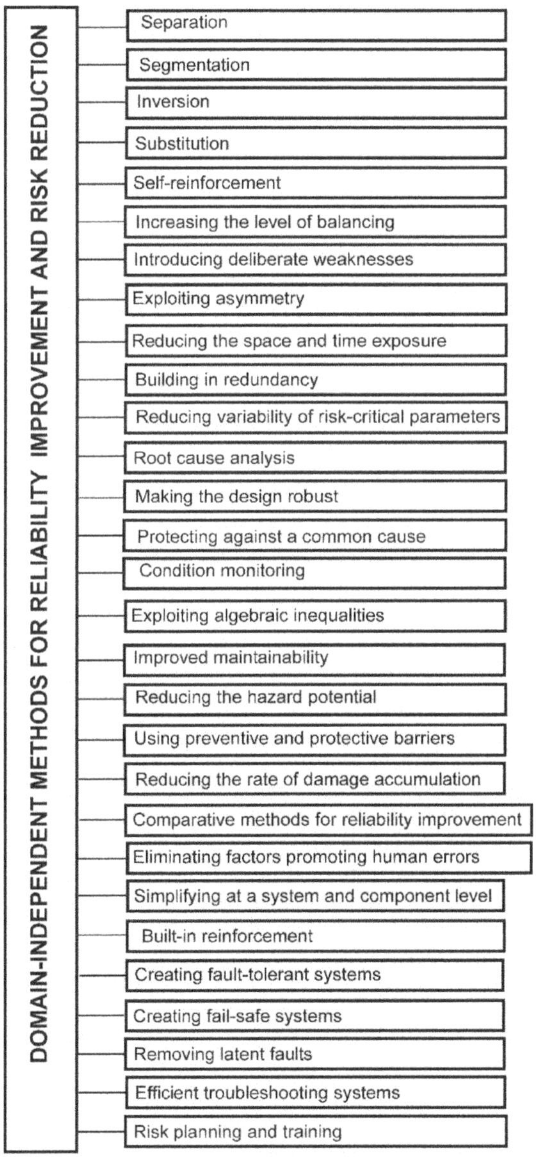

FIGURE 2.1 A list of domain-independent methods for reliability improvement and risk reduction.

Often, a reliability challenge in one domain has already been addressed in another through a similar method. For instance, replacing mechanical systems with electrical, magnetic, or optical ones to eliminate failure modes has been successfully applied in automotive engine control, temperature and stress measurement, and process control in manufacturing. The core idea is *substitution*, which involves replacing systems based on one physical principle with systems based on a different principle to eliminate inherent failure modes. This is a powerful, generic method applicable across many industries.

The domain-independent methods *provide a structured way of thinking* about reliability improvement and help prevent the continual reinvention of existing solutions. Many of these methods also promote *simple, low-cost solutions* in contrast to traditional approaches that rely on redundancy, condition monitoring, reinforcement, or expensive materials.

Many of the domain-independent methods shown in Figure 2.1 are discussed in Todinov (2019a). The categories outlined in this classification have been distilled from many engineering solutions, each of which was analysed for recurring risk reduction

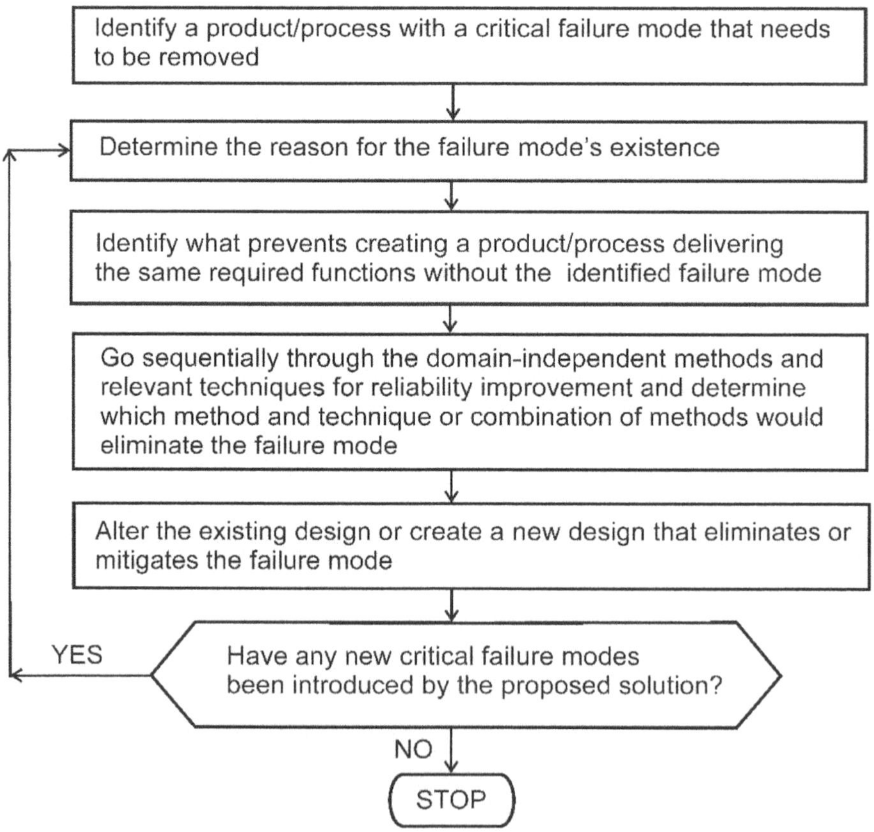

FIGURE 2.2 A general approach for reliability improvement by domain-independent methods.

patterns. The purpose of this book is also to introduce new domain-independent methods, along with classification of related techniques, case studies, worked examples and tutorials on reliability improvement. The general approach to be followed in reliability improvement by domain-independent methods has been given in Figure 2.2.

2.7 REQUIREMENTS FOR THE PROPOSED SOLUTION

The proposed reliability-improving solution must aim to improve the reliability of a selected component, assembly, or system or to reduce the consequences of its failure by applying one or more of the domain-independent methods listed.

The following requirements must be fulfilled:

Identification of Failure Modes:
The failure modes of the selected product or process must be clearly identified and explained. It should be demonstrated why these failure modes are critical and how they impact the reliability or performance of the system. A thorough analysis must ensure that no high-impact failure modes have been overlooked.

Demonstration of Reliability Improvement:
Strong, well-reasoned arguments must be presented to show that the proposed solution effectively eliminates or mitigates the identified failure modes. Evidence must be provided—through analysis, case examples, or calculations—that the reliability of the component, assembly, or system has indeed been improved, or that the consequences of failure have been reduced.

Justification of the Proposed Design:
The proposed design or redesign must be logically justified. This includes providing supporting analyses, calculations, and reasoning that clearly demonstrate the reliability gains.

Cost-Benefit Consideration:
It must be shown that the benefits of the proposed solution outweigh its cost. Ideally, the solution should result in improved reliability or reduced risk *without* significant increases in cost, mass, energy consumption, or spatial requirements. However, solutions that result in a modest cost increase may be acceptable if the anticipated reliability benefits significantly outweigh the added cost.

Environmental Impact:
The proposed solution must not have a negative impact on the environment.

Avoidance of New Failure Modes:
The modified or new design must be carefully analysed to ensure that it does not introduce additional critical failure modes.

In summary, the proposed solution must be technically sound, economically justified, environmentally responsible, and demonstrably effective in improving reliability or reducing failure consequences.

2.8 OVERCOMING MENTAL INERTIA

The greatest challenge in arriving at effective reliability-improving solutions is overcoming mental inertia. Mental inertia inhibits unconventional thinking and creativity. It often stems from conventional wisdom, tradition, and deeply held beliefs. Ironically, deep expertise in a specific field can also contribute to mental inertia. Experts may resist novel ideas within their domain because of their reliance on familiar patterns of thought. Moreover, the pursuit of complex, ambitious solutions can sometimes blind them to simpler, more obvious alternatives. Actively challenging mental inertia opens the door to more innovative and effective solutions.

Mental inertia manifests in several forms:

- *Familiar-area-induced inertia*: the tendency to cling to what is familiar and resist taking a new route;
- *Terminological inertia*: the rigid attachment to specialized language and jargon;
- *Perceptual inertia*: the fixed perceptions of what an object, system or process should look like;
- *Functional inertia*: the assumptions about what a product or process is supposed to do;
- *Cognitive inertia*: deeply rooted beliefs and assumptions;
- *Authority-induced inertia*: the uncritical acceptance of established theories, codes, or experts.

Fortunately, there are practical and effective strategies to overcome mental inertia. Here are some of them:

- *Avoid specialized terminology when framing a problem*; use general terms like *thing*, *object*, or *element*.
- *Reformulate the problem from a different perspective to uncover new possibilities*; invert the problem to gain fresh insight or challenge hidden assumptions. Working backward can bypass habitual thinking paths. Start from the desired outcome and reverse the steps.
- *Question established beliefs*, principles, and design norms, even those considered foundational.
- *Challenge assumptions*; hidden assumptions can severely limit innovation. List and question every assumption made about the problem. Prompt: 'What if this assumption were not true?' Use a 'Beginner's Mind' approach. Pretend that nothing is known about the problem domain. Ask basic, even 'naive' questions to expose unquestioned assumptions.
- *Use lateral thinking*. Employ Edward de Bono's techniques like random word association or provocation. Deliberately introduce illogical or absurd ideas to disrupt linear thinking.
- *Break functional fixedness*. Question standard functions of components or tools.
- Ask: 'What else can this object/system do?' Use analogies from different disciplines to inspire novel uses.

- *Employ morphological analysis*. Break down the system into functional units and explore combinations. This is useful for generating a wide range of design alternatives systematically.
- *Use analogical thinking*. Look for solutions from other industries.
- *Use cross-disciplinary collaboration*. Different disciplines approach problems with different frameworks. Involve professionals from other fields (e.g., software engineers in mechanical problems, or psychologists in human-factors engineering).
- *Use the SCAMPER technique* which is a structured creativity method that modifies existing ideas. Scamper is an acronym where the letters stand for:

  Substitute: What can be replaced?
  Combine: What ideas or processes can be merged?
  Adapt: What else is like this?
  Modify: Can it be magnified/minimized?
  Put to other use: Can it serve another function?
  Eliminate: What can be removed?
  Rearrange/Reverse: Can order or direction be changed?

- *List all available resources*. Think how these can be used in solving the problem.
- *Think of the ideal solution*. Think about how it can be achieved by using the available resources.
- *Employ constraint relaxation*. Temporarily relax or remove design constraints to explore broader solution spaces. Afterwards, reintroduce constraints to shape feasible options.
- *Intentionally seek dissonance or conflict*. Bring in 'devil's advocates' or dissenters to challenge consensus.
- *Alternate between convergent and divergent thinking*. Diverge to explore options without judgment then converge to select and refine. Balancing these phases prevents early lock-in to stale ideas.
- *Explore hidden variables or unmeasured influences*. Ask: 'What factors are not considered or measured?' Hidden variables often lead to unexpected root causes or leverage points.
- *Practice mindfulness and mental breaks*. Stepping away from the problem often allows the subconscious mind to work on solutions. Mindful activities reduce stress and increase mental agility.

3 Improving Reliability by Substitution

3.1 IMPROVING RELIABILITY BY SUBSTITUTION: CLASSIFICATION OF TECHNIQUES FOR THE METHOD OF SUBSTITUTION

Underlying Idea: The core concept behind the method of substitution is to eliminate failure modes in a system operating on one physical principle by replacing it with a system based on a different physical principle.

The classification of substitution techniques (see Figure 3.1) has been presented in Todinov (2022a) and includes the following categories:

- *Substitution with mechanical assemblies*
- *Substitution with electrical assemblies*
- *Substitution with magnetic assemblies*
- *Substitution with assemblies exploiting electromagnetic fields*
- *Substitution with assemblies exploiting electrostatic fields*
- *Substitution with acoustic assemblies*
- *Substitution with software-based assemblies*

Substitution can also be made by using a combination of the listed basic techniques. Substitution is justified only when it eliminates existing failure modes without introducing new, potentially more hazardous ones. While replacing a component to eliminate specific critical failure modes can be effective, it may also give rise to other significant vulnerabilities. For example, designs that involve substitution with electrical or magnetic assemblies should be carefully evaluated to ensure that no new critical failure modes are introduced.

3.2 EXAMPLES OF SUBSTITUTION ACROSS VARIOUS DOMAINS

3.2.1 Substitution with Mechanical Assemblies

In certain applications, replacing electrical or electronic devices with mechanical assemblies can significantly improve system reliability. Many electronic systems depend on consistent battery power and reliable electrical sources. However, they are

DOI: 10.1201/9781003679301-3

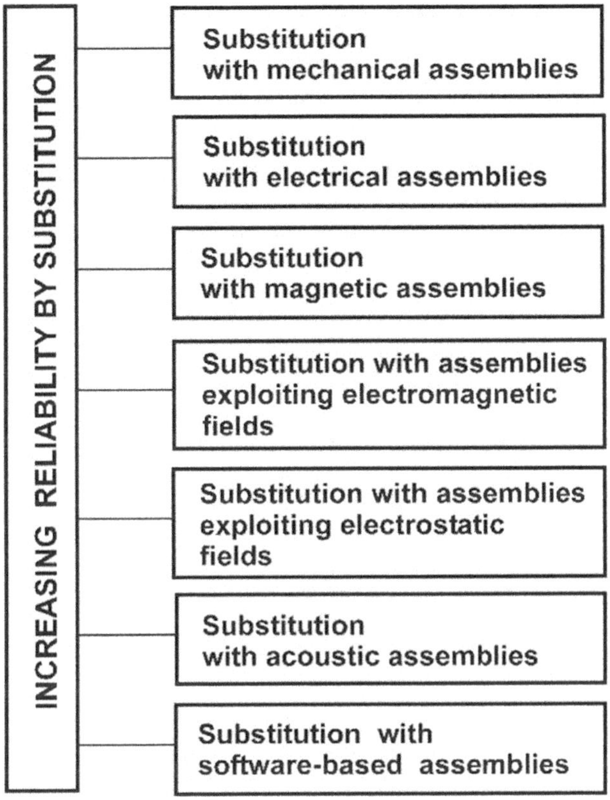

FIGURE 3.1 Classification of techniques for improving reliability by substitution.

often prone to risks such as overheating, fire, and failure due to environmental factors. Electronic components are particularly vulnerable in harsh conditions, such as high temperatures, strong electromagnetic interference, excessive vibrations, or elevated humidity.

When such environmental factors are present and the risk of failure is considered unacceptable, mechanical assemblies may offer a viable and more robust alternative. Mechanical systems are generally more resilient to electromagnetic radiation, extreme temperatures, moisture, and physical shocks or vibration.

One common example of this substitution is replacing battery-powered devices with spring-powered devices, thereby eliminating the risk of battery failure. A well-known instance is the use of spring-powered radios in remote locations where batteries are unavailable or impractical.

The use of mechanical interlocks is another example. Unlike their electrical counterparts, mechanical interlocks do not depend on sensors or wiring, making them more reliable in safety-critical applications. For example, machine guards often employ mechanical interlocks that physically prevent operation unless safety covers are properly secured.

Similarly, mechanical fuses or shear pins can replace electronic protection circuits. These mechanical, deliberately weakened points are simple, always operational, and immune to the failure modes typical of electronic assemblies.

Another effective substitution with mechanical systems involves replacing solenoid actuators with spring-loaded or gravity-driven mechanisms. These alternatives reduce dependence on electrical power and control systems, which can fail under various conditions. In emergency systems, for instance, gravity-based shut-off valves are often preferred over solenoid valves due to their inherent reliability and independence from electrical power.

By carefully selecting mechanical alternatives where appropriate, designers can enhance system durability and operational safety, especially in environments where electronic assemblies are more likely to fail.

3.2.2 Substitution with Electrical Assemblies

Replacing mechanical systems with electrical systems that deliver the same functions offers several important benefits:

- Reduces the amount of material required, contributing to more sustainable and environmentally friendly solutions.
- Lowers overall design complexity, making systems easier to develop and integrate.
- Minimizes the number of moving components, reducing potential points of failure.
- Electrical systems are less prone to physical degradation over time.
- Eliminates friction-based wear mechanisms common in mechanical systems.
- Enables more accurate and consistent operation.
- Simplifies maintenance procedures and reduces maintenance frequency.

Substitution with electrical or electronic assemblies has led to significant improvements in performance, reliability, and longevity across various applications. Well-known examples include:

- *Electronic control modules* substituting electromechanical systems for managing ignition timing in internal combustion engines. These offer greater precision and adaptability than electromechanical systems.
- *Replacement of mechanical relays* with solid-state relays, which eliminate mechanical wear and provide faster, more reliable switching.
- *Replacement of mechanical watches* with quartz watches greatly enhances timekeeping accuracy and reduces maintenance.
- *Replacement of mechanical pushbutton switches* (rated for tens of thousands of actuations) with *Hall-effect-based switches*, capable of tens of millions of actuations, provides vastly superior durability and resistance to environmental factors.
- Replacement of the mechanical controls of automobile braking and steering systems with 'by-wire' technologies. These systems can stop a car in an

emergency within a shorter distance compared to traditional mechanical systems.

3.2.3 Case Study: Detecting Through-Cracks in Protective Coating

Consider a test for detecting cracks in the protective coatings of steel cartridges used in corrosive environments (Figure 3.2). The plastic coating is essential for preventing corrosion, ensuring that no through-cracks are present. A traditional mechanical test, such as covering the cartridge with fluorescent paint, can reveal all cracks, including those that fully penetrate the coating. However, this method is associated with a serious drawback: it also highlights blind cracks that do not penetrate through the coating and therefore do not compromise corrosion protection (Figure 3.2). This may lead to unnecessary false alarm and costly overcorrection.

The electrical system shown in Figure 3.3 includes a battery (1) connected to two electrodes (2 and 3). Electrode (2) is immersed in an electrolyte solution (4) contained within a vessel (5). The electrolyte (4) is in contact with the outer surface of the cartridge (6). Electrode (3) is affixed to the inner surface of the cartridge (6). When a voltage is applied across the electrodes, the presence of a weak current, detected by the galvanometer (7), indicates the existence of through-cracks that penetrate the plastic coating. In the substituting electrical assembly, blind cracks no longer interfere with the test results or trigger false alarms.

In this way, the false indication by the mechanical, fluorescent paint method is avoided by substituting the mechanical testing assembly with an electrical testing assembly.

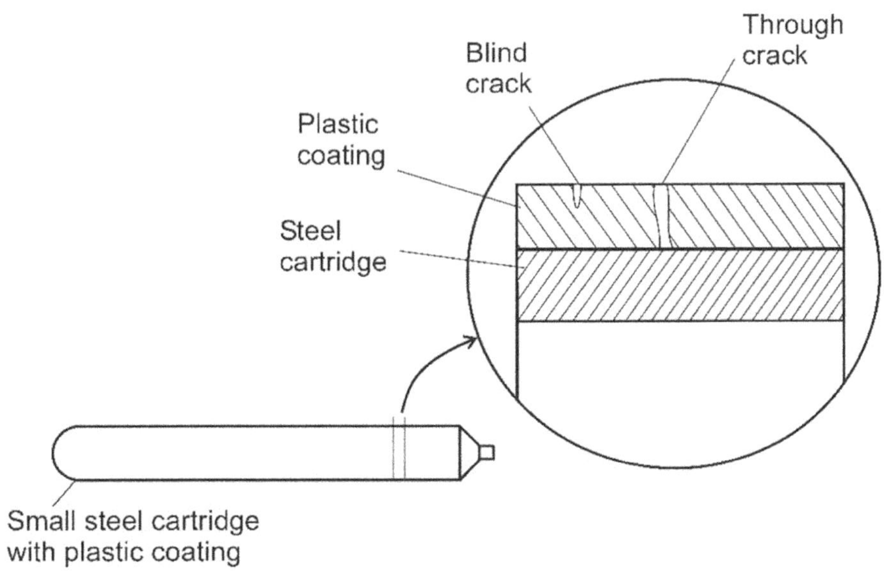

FIGURE 3.2 A small steel cartridge with blind cracks and through-cracks in the plastic coating.

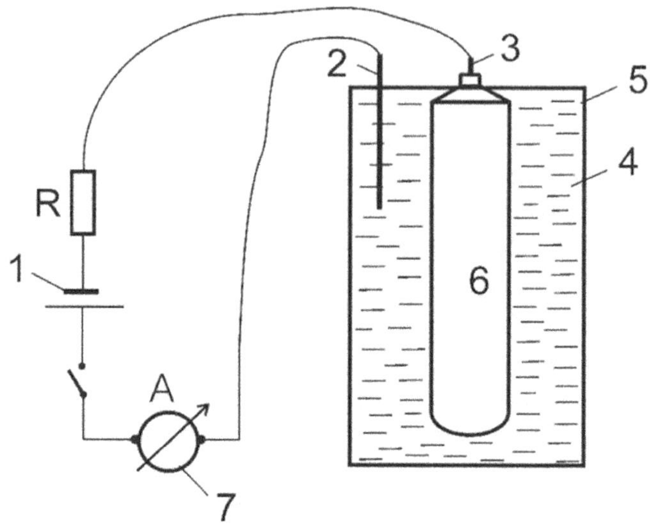

FIGURE 3.3 An electrical system used to detect the presence of through-cracks in the plastic coating.

3.2.4 Substitution with Magnetic Assemblies

The core concept behind the substitution with magnetic assemblies is to eliminate failure modes by replacing conventional systems with systems that operate on magnetic principles. A typical example is the magnetic stirrer, which eliminates the need for a mechanical seal. As a result, it removes failure modes associated with seal leakage and corrosion. It is a sustainable solution because it reduces the cost of materials and there are no leaks, hence environmental pollution is eliminated.

The magnetic couplings enhance operational safety in handling hazardous fluids. Thus, substituting shaft seals in chemical pumps with magnetic couplings prevents toxic leaks.

The use of magnetic levitation (Maglev) is another powerful application in transportation and for laboratory equipment. Magnetic bearings instead of conventional contact-based bearings do not suffer friction and wear and have extended life and reliability. They also offer ultra-clean and vibration-free operation which improves performance in precision environments.

A related example is the magnetic worm gear drive assembly (Figure 3.4; patent US 3814962A). This design avoids several failure modes commonly found in traditional gear systems:

- There is no wear as torque is transferred without physical contact;
- There are no lubrication-related failures since lubrication is unnecessary;
- The maintenance is simplified, hence the risk of maintenance errors is significantly reduced;
- There are no misalignment failures due to non-contact operation;
- Vibration transmission is eliminated which enhances operational smoothness.

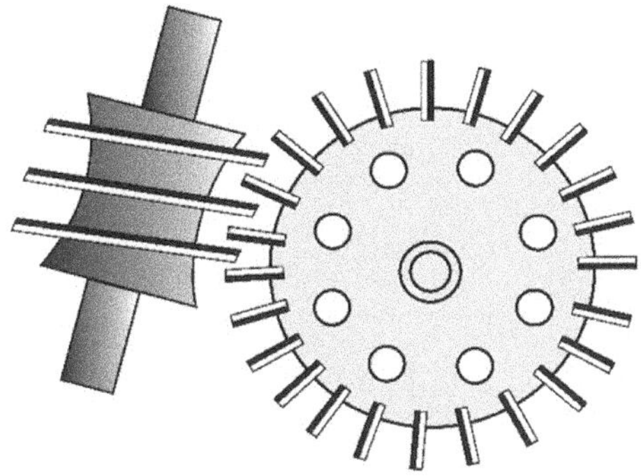

FIGURE 3.4 Magnetic worm gear drive assembly (patent: US 3814962A).

Consider a safety valve on a pipeline carrying flammable toxic fluid. The valve is maintained in an open position by hydraulic pressure. In case of a loss of pressure in the hydraulic line, the valve must be returned in closed position by a mechanical compression spring. Due to failure caused by fatigue, corrosion, or sagging of the spring (from stress relaxation), the spring may fail to return the valve in closed position and stop the toxic fluid in the pipeline. This poses a significant safety risk which could be eliminated if the mechanical spring is replaced by a magnetic spring (Figure 3.5). Replacing spring-loaded with magnetic latches also prevents fatigue-related failures.

Substituting a mechanical spring with a magnetic spring eliminates several major failure modes inherent to mechanical springs, including *fatigue failure*, *sagging,* and *corrosion*, without introducing new failure modes.

A further example involves using reed switches in hazardous gas detectors instead of electromechanical relays. Replacing traditional relays with magnetic reed switches eliminates arcing and contact wear. This improves safety in explosive or corrosive environments and enhances reliability in high-cycle switching applications.

Another application of magnetic substitution is magnetic coupling (Figure 3.6). This type of coupling eliminates the transmission of noise and vibration from input shaft '1' to output shaft '2', as it operates without physical contact or friction. As a result, it does not suffer from wear-related degradation. Additionally, it is tolerant of both parallel and angular misalignment. Magnetic coupling enables torque transmission through a physical barrier and functions as a torque limiter, protecting input shaft '1' from overload if output shaft '2' encounters a sudden resistance.

One more effective application of this type of substitution is replacing mechanical attachments with magnetic attachments in situations where a quick release is required, such as in emergency conditions. A particularly interesting use of magnetic assemblies is their role as *deliberate weaknesses*. These are components designed to decouple two parts of a system when the load exceeds a specified threshold. In such cases, the magnetic assembly disengages under overload conditions, thereby releasing the load and preventing damage or catastrophic failure. Compared to mechanical deliberate weaknesses (e.g., shear pins, rupture disks), magnetic assemblies offer a significant advantage: they can be reset or restored quickly and at no cost, ensuring minimal downtime and improved system resilience.

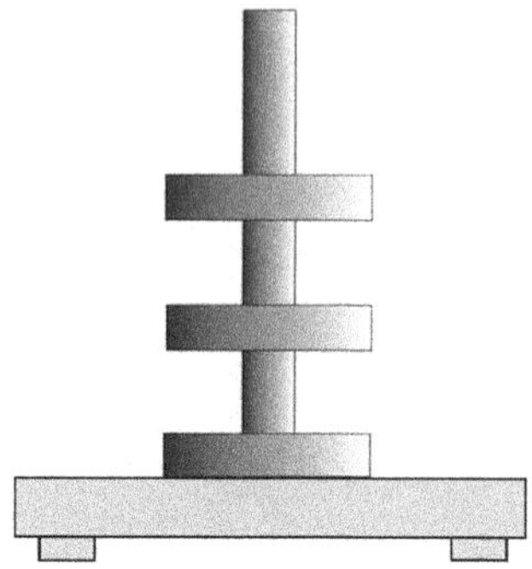

FIGURE 3.5 Magnetic spring.

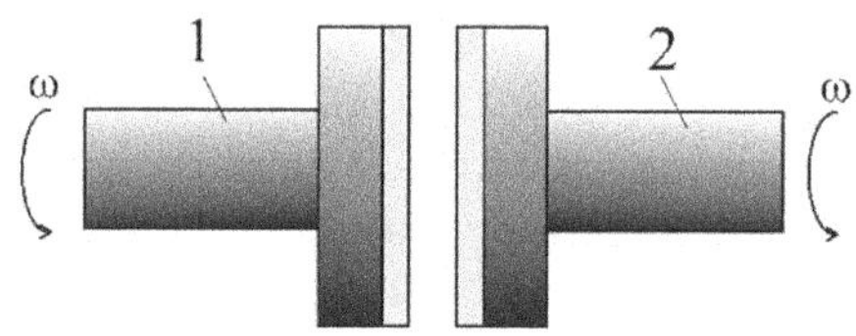

FIGURE 3.6 Magnetic coupling.

3.2.5 Substitution with Assemblies Exploiting Electromagnetic Fields

The most common substitution of this type is the substitution with optical systems which offers numerous advantages: it simplifies design, eliminates moving parts, reduces wear and fatigue, minimizes accumulated damage, increases precision, and enhances maintainability.

A well-known example of substitution with optical assemblies is the replacement of the electro-mechanical computer mouse (using a rolling ball) with an optical mouse. Compared to its mechanical counterpart, the optical mouse provides:

- Significantly improved durability;
- No accumulation of dirt or debris;
- Easy maintenance with no cleaning required;
- Operation on a wide variety of surfaces, including glass;
- Enhanced motion tracking;
- Higher precision.

Another impactful application of a substitution of this type is the substitution of electrical wiring with fibre optic lines, which results in:

- Superior performance in harsh environments (e.g., high humidity, pressure, voltage, or temperature);
- Complete immunity to electromagnetic interference;
- Reduced long-term degradation and damage accumulation.

The optical proximity sensors versus mechanical limit switches is yet another example. The mechanical switches detecting presence of objects are substituted with infrared or laser-based proximity sensors. The reliability improvement is due to the absence of mechanical contact. This leads to less wear and contamination which makes the optical sensors more durable in corrosive environments. In addition, there is a faster response and higher precision.

Substitution with optical systems is also *highly effective in eliminating failure modes in measurement and testing*. A well-known example is the use of an infrared optical thermometer in place of a thermocouple. This substitution addresses several critical failure modes of thermocouples:

- Measurement variability due to inconsistent thermocouple attachment;
- Disruption of the temperature field and errors caused by the physical contact;
- Rapid degradation of the mechanical contact, reducing durability;
- Limited measurement speed;
- Low reliability at temperatures above 1300 °C;
- Inaccurate measurements on high-voltage surfaces due to electrical interference.

Using Wi-Fi or Bluetooth instead of wired connections can offer several reliability advantages. Here's a breakdown of the key reliability benefits:

- There are no connectors and cables to break, fray, or get unplugged accidentally.
- Reduced wear and tear, which is important in environments with movement or vibrations (e.g., robots, portable devices).
- Provide better resilience to physical damage or obstructions because there is no dependency on ports or connectors, which can wear out or get damaged.
- Wireless links can still work even if the device is in a protective enclosure where cables would be impractical. As a result, wireless connections are more robust in industrial, outdoor, or harsh environments.
- Most modern Wi-Fi and Bluetooth devices automatically reconnect if the signal drops.
- Wireless connections eliminate single point of cable or connector failure. A single damaged wire or connector in a wired system can disrupt the entire connection while wireless communication is not subject to mechanical damage. The result is improved overall system uptime and reduced maintenance needs.
- The wireless connection has a simplified maintenance and troubleshooting which makes it possible to move or swap devices without worrying about cabling. There are also fewer components to inspect during troubleshooting.

A final example is the replacement of mechanical buttons and switches with optical touchscreens. This offers the advantage of eliminating mechanical contacts that can wear out over time, while also providing sealed surfaces that protect against contaminants.

3.2.6 Substitution with Assemblies Exploiting Electrostatic Fields

Electrostatic field-based substitution can also be employed to improve reliability.

In particular, electrostatic precipitation (ESP) provides several advantages over mechanical filtration for removing particulate matter from gas streams. The key benefits include:

- *Superior particle removal efficiency*: ESP units can capture extremely fine particles, which are typically beyond the capabilities of mechanical filters. This makes them especially effective for submicron particles commonly found in smoke and industrial emissions.
- *Lower energy consumption*: ESP units generate significantly lower pressure drops compared to mechanical filters. As a result, less energy is required to move air or gas through the system. Mechanical filters, by contrast, become increasingly resistant to airflow as they accumulate particles, thus raising energy costs.
- *Extended operational life*: ESP units can function for long periods without frequent cleaning or replacement. Mechanical filters, on the other hand, tend to clog over time, necessitating regular maintenance and leading to higher operational costs.
- *Cost-effectiveness for large-scale applications*: In industrial settings that handle high volumes of gas, ESP units offer long-term cost advantages due to reduced energy and maintenance needs. Mechanical filters are often less scalable, requiring multiple or larger units to achieve similar performance.
- *Adaptability to challenging environments*: Wet ESP units are particularly suitable for gas streams that are humid or sticky. These are conditions under which many mechanical filters perform poorly or fail entirely.
- *High-temperature and corrosive gas resistance*: ESPs can operate effectively in harsh environments involving elevated temperatures or corrosive gases, where mechanical filters are likely to degrade or become unusable.

Electrostatic fields play a significant role in various advanced technologies, including soft robotics and materials science. In soft robotics, electrostatic fields are harnessed to create electrostatic artificial muscles. These actuators offer a lightweight, flexible alternative to traditional pneumatic or motor-driven systems, enabling smoother and more precise movements. By eliminating the need for bulky components, electrostatic actuators contribute to the development of more compact, efficient, and responsive robotic systems.

In the field of materials research, electrostatic levitation is employed as a contactless method for suspending high-temperature samples. This technique replaces conventional mechanical holders, which can introduce contamination or interfere with the experiment due to thermal or mechanical constraints. By levitating the sample using controlled electrostatic forces, researchers can study phenomena such as crystal growth during melting and solidification under highly pure and stable conditions. This is particularly valuable for investigating materials under extreme thermal conditions, where traditional support structures would fail.

3.2.7 Substitution with Acoustic Assemblies

Improving the reliability of mechanical or electrical assemblies by substituting them with acoustic assemblies is a strategy that leverages sound-based mechanisms (e.g., ultrasonic waves, surface acoustic waves or audio signals) to reduce moving parts, contact wear, or electromagnetic interference which entail various failure modes. Below are examples where such substitution has led or can lead to improved reliability.

Ultrasonic sensors replace mechanical limit switches or contact-based sensors. This substitution eliminates moving parts and with them – failure modes associated with wear, misalignment, and mechanical failure. Ultrasonic sensors are also resistant to dust, humidity, and vibrations.

An electro-mechanical system for measuring the liquid level in a reservoir can also be substituted with an acoustic measurement system by measuring the resonance frequency of the reservoir. The resonance frequency is different for different levels of the liquid in the reservoir.

Acoustic flow measurement can substitute mechanical flow meters. This substitution uses ultrasonic flow meters instead of paddle wheel or turbine flow meters.

Because of the absence of moving parts, there is less maintenance and failures associated with incorrect maintenance. Acoustic flow meters work well with corrosive or particulate-filled fluids and do not suffer from failure modes associated with corrosion and erosion and maintain accuracy over longer periods.

Acoustic levitation can also be used instead of mechanical conveyors. The core idea is using acoustic levitation to move particles/droplets instead of micromechanical systems. The advantage is that friction, contamination, and mechanical wear are eliminated. This technique is ideal for sterile environments or sensitive biological samples.

3.2.8 Substitution with Software-Based Assemblies

Reliability can be significantly improved through the substitution of mechanical systems with software-based systems. For instance, complex kinematic functions that traditionally require intricate mechanical components can be implemented via software in conjunction with servomotors. This shift reduces mechanical complexity, enhances design flexibility, and significantly reduces physical wear, which is one of the primary causes of unreliability. Furthermore, software replication introduces no manufacturing variability, ensuring consistent performance across units.

Software integration also enables intelligent features that enhance both safety and usability. Real-time guidance, warnings, and user feedback can help prevent hazardous actions, reduce injury risk, and avert operational failures. Additionally, software-based diagnostic tools facilitate rapid troubleshooting and corrective actions, mitigating the severity and duration of failures.

Replacing mechanical assemblies with software and electronic assemblies allows for the seamless integration of sensing and adaptive functionalities. These systems can autonomously reset objectives and adjust to dynamic environmental conditions, thereby increasing resilience and overall functionality. For example, modern air conditioning systems utilize sensors to measure both temperature and humidity, adapting their output through fuzzy logic algorithms. This not only improves comfort but also extends system longevity.

A more sophisticated example is the programmable electronic control unit (ECU) found in contemporary vehicles. ECUs govern multiple engine parameters, including valve timing, ignition timing, transient fuelling, air/fuel ratios, and cold-start corrections. These software-driven controls enhance engine reliability, reduce fuel consumption and emissions, and contribute to longer engine life.

Software also underpins advanced technologies such as artificial intelligence. Autonomous vehicles, for example, rely on software to process sensor data, recognize traffic lanes, detect obstacles and pedestrians, plan routes, and execute safe driving manoeuvres. These capabilities mitigate risks associated with human limitations such as fatigue, distraction, or slow reaction times.

Supporting technologies in software-based visual recognition systems include machine vision, artificial neural networks, and deep neural networks. These tools enable fast and accurate object detection and classification, further enhancing the safety, performance, and reliability of complex systems.

4 Improving Reliability and Reducing Risk by Separation

4.1 THE METHOD OF SEPARATION: CLASSIFICATION OF TECHNIQUES RELATED TO THE METHOD OF SEPARATION

This chapter introduces an important, domain-independent method for improving reliability and reducing risk referred to as *the method of separation.*

Failures often result from the harmful interaction of factors critical to reliability and risk. Therefore, separating these risk-critical factors to reduce their interactions represents a powerful path to enhancing reliability and mitigating risk.

The core idea of the method of separation is to eliminate failure modes by separating risk-critical factors, functions, properties, and geometry.

Despite its relevance, the method of separation has not been explicitly discussed in the literature as a dedicated risk-reduction strategy. Interestingly, many established risk-reduction techniques in engineering are in fact applications of this method, but they have not been recognized or categorized as such.

For example, the method of division of tasks presented by Pahl et al. (2007) is essentially an application of the method of separation. Similarly, the widely used concept of barriers, as discussed by Svenson (1991), Leveson (2011), Hollnagel (2016), and Eder and Hosnedl (2008), also reflects the principle of separation.

Barriers that distance triggers from hazards reduce the likelihood of accidents, while those that separate hazards from potential targets reduce the consequences after the accidents occur. The method of separation has also been featured in TRIZ (Altshuller, 1984, 1996, 1999; Terninko et al., 1998; Salamatov, 1999; Savransky, 2000) as a powerful methodology for inventive problem solving, where it is used to resolve physical contradictions, for example: "an object must have attribute *A* during one phase of use and not-*A* during another." However, TRIZ applies separation primarily for innovation, not for reliability improvement or risk reduction. It does not address the mechanisms by which separation enhances reliability, nor does it provide guidance on the systematic implementation of this method in design for reliability. More recent TRIZ-oriented texts (Orloff, 2006, 2017; Rantanen and Domb, 2008; Gadd, 2011; Petrov, 2020; Starovoytova, 2021) also lack any treatment of these mechanisms.

Furthermore, TRIZ does not account for *stochastic separation*, in which separation occurs with a certain probability, nor does it consider *logical separation*, where

DOI: 10.1201/9781003679301-4

interaction is prevented without relying on time, space, or conditional separation. These types of separation are significant, as both types play important roles in risk mitigation.

It is also crucial to emphasize that the method of separation is not equivalent to the traditional use of *barriers separating hazards from targets*. While barriers are a special case of separation and have seen widespread application in risk management (Haddon, 1973; Hale et al., 2004), the method of separation encompasses a broader range of mechanisms, many of which remain unknown or underutilized by engineers, designers, and reliability professionals.

Examples include:

- *Stochastic separation*, where the interaction between risk-critical factors is probabilistically prevented;

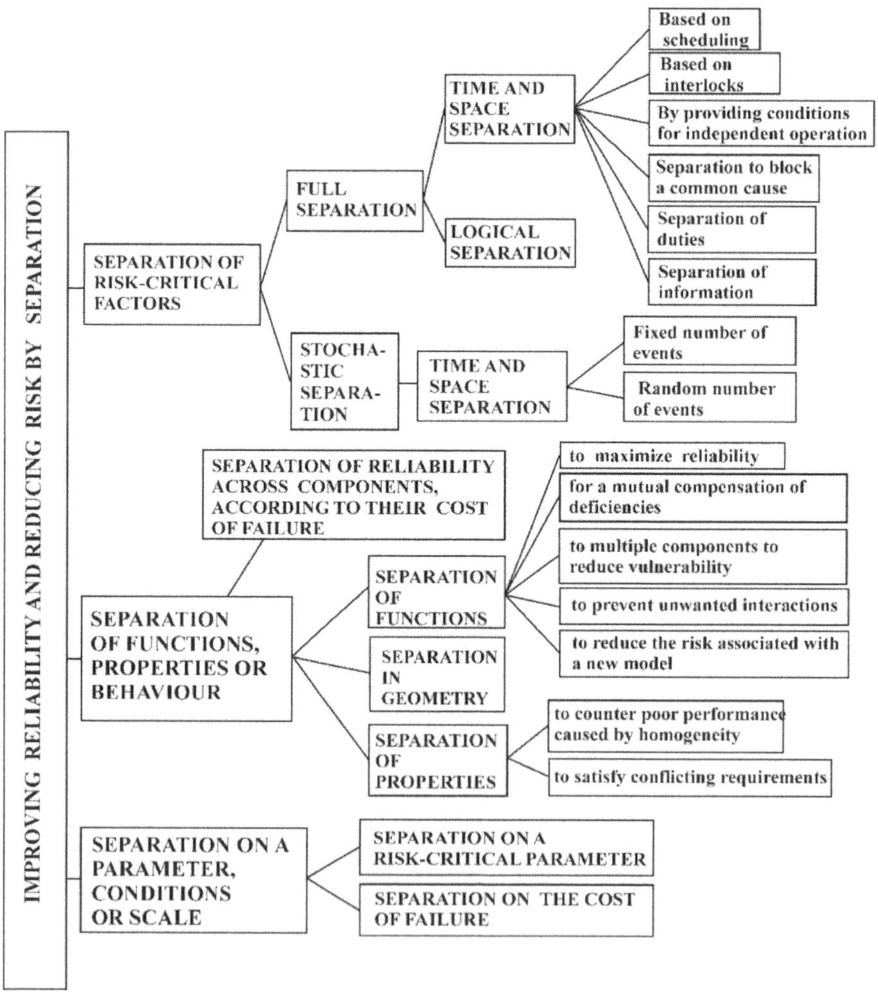

FIGURE 4.1 Classification of different separation mechanisms and techniques.

- *Separation based on the cost of failure* is a largely overlooked strategy despite the prevalence of high-consequence failures where component reliability does not match the associated risk;
- *Separation of properties to counteract performance degradation due to homogeneity* is a concept often overlooked in engineering design.

A recurring limitation in the existing literature is the absence of discussion on the mechanisms through which separation improves reliability and reduces risk. This knowledge gap hinders the systematic application of this method in engineering practice.

Figure 4.1 summarizes the classification of separation techniques, serving as a foundation for their structured implementation in design processes (Todinov, 2017a).

4.2 SEPARATION OF RISK-CRITICAL FACTORS

Separation of risk-critical factors is essential when scheduling critical random events where any overlap could lead to severe consequences. In such cases, the simultaneous occurrence of these events must be entirely prevented.

Common examples of separation of risk-critical factors include:

- *Time-based separation*, as seen in the use of traffic lights, which prevent collisions by regulating the movement of intersecting traffic flows and pedestrian crossings.
- *Space-based separation*, such as grade-separated intersections that isolate pedestrian and vehicular traffic at different levels, thereby eliminating the possibility of direct interaction and reducing the risk of accidents.
- *Separation of risk-critical factors by using a barrier* is used to distance risk critical factors such as high-temperature, abrasive action of particles suspended in fluid, vibrations, and humidity.

A well-known example of separation of risk-critical factors is mitigating vibrational fatigue. Early rotor systems in helicopters relied on metallic bearings and hinges, which transmitted intense vibrational loads from the rotating blades directly to the fuselage. These vibrations led to fatigue in various components, necessitating frequent inspections and premature part replacements due to vibrational fatigue failures.

The solution involved isolating vibrational loads by incorporating elastomeric bushings into the main rotor hub and control linkages. These bushings, composed of multi-layer rubber-steel laminates, were capable of deforming under load while effectively damping vibration energy. This design allowed for significant vibration isolation, drastically reducing vibrational loads and the associated fatigue.

In industrial slurry-handling systems, pumps often draw in suspended sand, silt, and other abrasive particles that can erode impellers, seals, and casings. To improve reliability, a cyclone separator (hydrocyclone) can be installed.

The incoming fluid enters the hydrocyclone tangentially, forming a high-velocity vortex. Centrifugal forces drive heavier abrasive particles toward the outer wall, where they are directed to the underflow for collection or discharge. The cleaner, particle-reduced fluid exits through the overflow and is supplied to the pump.

This arrangement reduces wear on impellers and mechanical seals, extending pump life and lowering maintenance frequency.

Safety fuel cells, constructed with a tough, flexible lining enclosed within a rigid steel or aluminium container, offer another example of separating risk-critical factors. The flexible, high-strength lining is engineered to deform on impact rather than rupture or fragment, helping to keep the fuel contained and significantly reducing the risk of fire in a collision.

4.2.1 Time Separation Based on Scheduling

Full separation in time is required from risk-critical factors or entities whose simultaneous presence must be excluded completely.

The time separation by scheduling enforces consistent time spacing between hazardous events. The full time separation is used in air traffic control, where it enforces consistent time spacing between arriving aircrafts, based on real-time information about the weather, headwinds, altitude, and speed. In this respect, the air-traffic control clearance is an important instrument providing the necessary separation to avoid collisions within controlled airspace. The separation provided by the air-traffic control clearance also guarantees a sufficient runway approach capacity which also keeps the risk of accidents low.

Another example can be given with time separation of service support into different time zones. This can be used for a vital continuous 24/7 maintenance support, that guarantees the safe operation of many systems.

4.2.2 Time and Space Separation by Using Interlocks

To prevent the simultaneous occurrence of two events, a *sentinel mechanism* can be implemented. This sentinel maintains a state, either *busy* or *free*. If an event is currently active, the sentinel returns a *busy* state, thereby blocking the initiation of any new event. A new event can only begin when the sentinel state is *free*. Once an event is completed, the sentinel's state is reset to *free*. This logic can also be integrated into an *interlock system*, effectively preventing any overlap between events.

Physical space separation can also be achieved through mechanical or electrical interlocks. One such method involves *disabling a power or control circuit* to ensure safe operation. A common example is found in rotating machinery: removing the protective shield interrupts the power supply, rendering the machine inoperable until the shield is securely repositioned. This greatly reduces the risk of injury by ensuring safe conditions before operation. Thus, to prevent microwave radiation leakage, microwave oven will not operate unless the door is fully closed. Another common example is in aviation where a landing gear cannot retract unless full weight is off wheels. This interlock prevents collapsing gear on ground. Interlocks are also commonly used for prevention of storage tanks' overfill by connecting them with flow cutoff systems.

A more advanced form of spatial interlock is the *presence-sensing safeguard system*, which halts the operation of hazardous equipment when a person enters a danger zone. These systems often use laser beams, infrared sensors, or light curtains. A light curtain emits a grid of beams; if any beam is obstructed, indicating a person's presence, the control circuit is opened and the power to the hazardous equipment is immediately shut off.

Another effective technique is *the trapped-key interlock system*. In this setup, access to hazardous equipment for maintenance is controlled by a key that remains trapped in the access door until the door is securely closed. Opening the door with the key simultaneously actuates a switch that interrupts the power supply. The equipment cannot be re-energized until the door is fully closed and the key is returned to its original position, ensuring that the system cannot be operated unsafely.

4.2.3 Logical Separation by Using a Shared Unique Key

Logical separation occurs when it is logically impossible for a hazardous operation to take place at a specific time or location. It also applies when it is logically impossible for two or more objects, such as hazards, triggers, or targets, to be in dangerous proximity to one another at the same time or in the same space.

Unlike physical separation, *logical separation does not rely on physical barriers*. Instead, the system is structured in such a way that hazardous interactions are *inherently impossible by design*. The proximity of triggers to hazards, or hazards to potential targets, is prevented by the *logic governing the system's operation*.

State validation logic is a common technique for logical separation, where an operation is allowed only if the system is in a known safe state. Thus, to prevent dangerous chemical reactions, programmable logic controllers block the simultaneous opening of incompatible chemical valves.

A variation of this technique is the *interdependent control*, where one function is blocked unless another condition is satisfied. For example, to prevent arcing and equipment failure, generator synchronization logic blocks connection to the grid unless the phase, frequency, and voltage match.

Another technique for logical separation is the use of *permissions* or *authentication*, where only certain users or conditions can unlock actions. Thus, to prevent data loss or security breaches, software systems commonly implement *role-based access control*, which makes certain actions (e.g., 'delete') inaccessible to unauthorized users.

Logical separation can also be used to prevent the *dangerous simultaneous execution of incompatible processes*. One practical method involves a *shared unique key mechanism*: a single unique key is required to activate each process involved in a given task. Since the key cannot be in two places at once, *simultaneous activation is logically impossible*, thereby eliminating the risk of hazardous process overlap.

4.2.4 Case Study: Logical Separation of X-Ray Equipment by a Unique Shared Key

Suppose a particular task requires adjusting a specimen before activating an X-ray beam (Figure 4.2). A critical failure mode arises when someone in Room

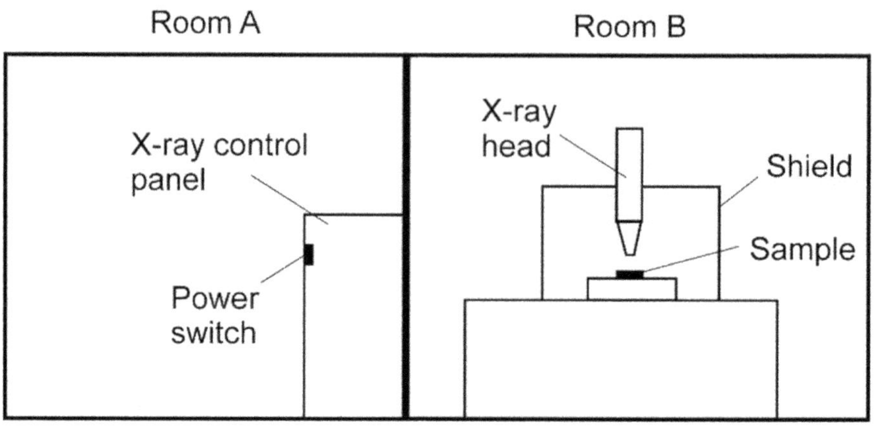

FIGURE 4.2 A critical failure mode: Someone switching on accidentally the X-ray equipment in Room A, while a person is adjusting the sample under the X-ray head in Room B.

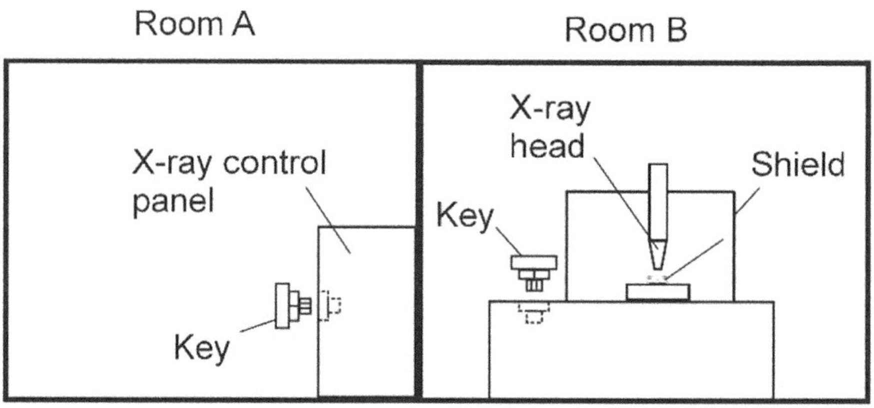

FIGURE 4.3 A solution based on logical separation by a unique shared key.

A accidentally switches on the X-ray equipment while another person is adjusting the specimen under the X-ray head in Room B. This failure mode exists because the specimen must be positioned under the X-ray head in Room B prior to beam activation.

To eliminate this failure mode, a logical separation of potentially dangerous actions is introduced using a unique shared key. This key is required both to unlock the protective shield for specimen adjustment in Room B and to activate the X-ray beam from Room A (Figure 4.3). Since the same key is needed in two physically separate locations, it is impossible to perform both actions simultaneously. This solution ensures that the X-ray cannot be activated while someone is adjusting the specimen, effectively preventing the dangerous scenario.

This approach uses logical separation to eliminate the safety risk in a simple, low-cost, and highly effective manner. Despite its simplicity and power, the use of a shared unique key is an underutilized method for improving safety and reducing risk in system design.

4.2.5 Space Separation of Information

This separation technique reduces the risk of information overload which is a common source of costly human errors. To mitigate errors caused by complexity and excessive information, machine-human interfaces are designed to present information in layers, displaying only what is immediately relevant to the task at hand.

A frequent design flaw in machine-human interfaces is presenting all available options to the user, which can lead to cognitive overload and increased likelihood of error. By applying information separation, the interface ensures that users only see options pertinent to their current needs; those irrelevant to their role or task remain hidden.

4.2.6 Separation of Duties to Reduce the Risk of Compromised Safety, Errors, and Fraud

The core principle of this separation technique lies in dividing a key task, particularly one vulnerable to misuse, into subtasks assigned to different individuals. This separation of duties enhances accountability, reduces conflicts of interest and human error, and ultimately lowers risk significantly. Failure to implement this principle properly can lead to serious consequences. For instance, a unit responsible for conducting hazardous operations should not be solely in charge of the associated safety controls. When operational units assess their own safety performance, the substantial effort and cost invested can result in confirmation bias, leading them to overstate safety and underestimate risks. By assigning safety oversight to an independent unit that reports directly to top management, the organization can greatly improve safety.

Another clear example arises in security system development. A person tasked with designing a security system must not also test it, as this creates a conflict of interest, requiring someone to evaluate vulnerabilities in their own work. This same principle is widely applied in software engineering, where separating the roles of software design and testing is essential to avoid costly errors. Combining these tasks increases the risk of undetected flaws due to personal bias or oversight.

In financial operations, separation of duties across roles, such as authorizing transactions, receiving and depositing funds, recording transactions, and reconciling statements, minimizes opportunities for fraud. No single individual should have the power to both perpetrate and conceal fraudulent activity. By ensuring that multiple people are involved, collusion becomes a necessary (and less likely) condition for fraud.

Similarly, in sales operations, separating the role of conducting sales from managing customer deliveries prevents fraud such as fictitious sales made to non-existent customers, which can be used to siphon company funds.

4.2.7 Separation to Block a Common Cause

A *common-cause failure* is usually due to a single cause with multiple failure effects that are not consequences of one another (Billinton and Allan, 1992).

A common cause reduces the reliability of multiple components simultaneously. The affected components are then more likely to fail, which reduces the overall system reliability.

Typical conditions that promote common-cause failures include common design faults, common manufacturing faults, common installation and assembly faults, common maintenance faults, and shared environmental stresses experienced by several components. Examples of such stresses are high temperature, pressure, humidity, erosion, corrosion, vibration, radiation, dust, electromagnetic interference, impacts, and shocks. A common cause may also arise from shared infrastructure, such as a common power supply, communication channels, or software.

For example, two programmable devices produced by different manufacturers and assembled and installed by different people can still suffer from a common cause if the same faulty piece of software has been installed on both devices. Maintenance and operational actions common to different components are also major sources of common-cause failures. Software routines written by the same person or team often exhibit similar faults.

Acceleration stresses leading to damage accumulation and rapid wear are typical examples of common causes. These include temperature, humidity, vibration, pressure, voltage, current, and the concentration of particular ions. This list is only a sample of possible acceleration stresses and can be extended. Acceleration stresses lead to faster wear; they increase the likelihood of failure for groups of components, thereby reducing overall system reliability.

A typical example of a common-cause is high temperature, which increases the susceptibility to deterioration of a group of electronic components. By simultaneously increasing the hazard rates of the affected components, the probability of system failure rises. Humidity, corrosion, or vibration also increases the joint probability of failure of affected components and shortens the system's lifespan. Even in systems with a high level of built-in redundancy, a common cause often leads to the failure of all redundant components within a short time, negating the benefit of redundancy.

Failure to account for acceleration stresses acting as common causes typically leads to overly optimistic reliability predictions; the actual reliability is much lower than predicted.

In many cases, the *separation principle* helps in blocking out common causes, thus reducing the risk of failure. Separating components by distances greater than the radius of influence of the common cause is an effective way to mitigate the risk of common-cause failures. Thus, multiple backups of vital information stored in different locations protect against data loss due to fire, theft, or sabotage.

In another example, clustering large fuel tanks within a single fuel depot (Figure 4.4a) increases the risk of cascading explosions if a fire occurs in any one of the tanks. In contrast, spacing the tanks at safe distances 'a' (Figure 4.4b) effectively prevents such cascading failures.

Likewise, positioning communication centers farther apart than the destruction radius of a missile increases the likelihood that some of the centers will survive the missile attack.

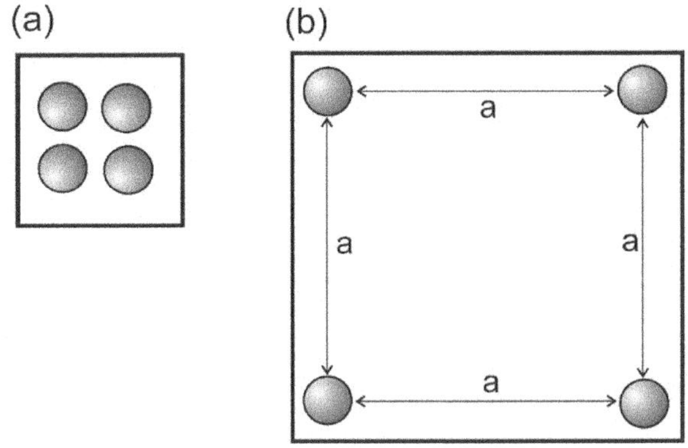

FIGURE 4.4 (a) Clustered fuel tanks in a fuel depo create the possibility of cascading failures; (b) spacing fuel tanks at safe distances.

Another implementation of this principle involves separating vital control components from components whose failure could cause damage. A typical example is routing control lines at safe distances from airplane jet engines. In the event of an engine explosion, the flight controls remain operational, allowing a safe landing. Separating components and insulating them from corrosive environments and environments with excessive dust, humidity, heat, or vibrations is also an effective way to guard against common-cause failure.

In subsea production stems, encapsulating electronic modules in elastomers and resins insulates them and protects against common-cause corrosion caused by marine water ingress.

Providing maintenance for redundant components by separate operators reduces the likelihood of common-cause failure due to faulty maintenance.

Design diversity consists of separating the physical principles on which redundant devices operate and is another efficient way to block common causes and reduce common-cause failures. The goal is to prevent multiple components from being affected by the same fault. For example, if two cooling pumps (a main pump and an emergency pump) cool a chemical reactor, failure of both creates an emergency. If the pumps come from different manufacturers or operate on different physical principles, common-cause faults can be avoided. If one is powered electrically and the other relies on gravity, the common cause of power failure is eliminated. Additionally, if they are maintained by different operators, the risk of failure due to a common faulty maintenance is reduced. Likewise, a common cause such as incorrect calibration of measuring instruments can be avoided if calibration is performed by different individuals.

Sundararajan (1991) suggests a *preliminary common-cause analysis*, which involves identifying all possible common causes to which a system is exposed and evaluating their potential effects. The purpose is to alert risk analysts to possible vulnerabilities.

Avoiding common links that can be impacted by a single cause is another effective strategy. Examples include avoiding a single location for critical components or a single data storage site. The destruction of all communication lines due to accident or terrorism can be prevented by not placing all lines in a single conduit.

A common-cause failure due to a software bug, for instance, can be prevented by providing an alternative algorithm and implementation for the same task, or by having a separate team independently develop the same piece of software.

Finally, *diversifying investments* across unrelated sectors protects against common-cause failures that could simultaneously reduce returns from multiple related sectors. Such examples are agricultural sectors affected by adverse weather or disease, consumer sectors impacted by a health scare, or multiple sectors within a country affected by political, economic, or social crises.

4.2.8 Separation by Providing Conditions for Independent Operation

The increasing complexity of modern technology has led to greater coupling and interdependence among components and subsystems. This elevated coupling means that the failure of one component is more likely to adversely affect others, potentially leading to cascading failures. When the failure of one component increases the likelihood of failure of another component, system reliability can be enhanced by ensuring that the components operate independently. Creating conditions for such independent operation is referred to as *full separation*.

For example, consider two devices where the second is powered by the first. If the first device fails, the second loses power. System reliability can be improved by separating their power sources, ensuring that the failure of the first device does not impair the operation of the other.

To reduce costs, devices often share control or communication channels. Consider a dual control system in which two command centres CC1 and CC2 control an electromechanical device *M* (Figure 4.5). In a cost-saving configuration (Figure 4.5a), both command centres share a single control channel. However, this setup lacks independence; failure of the shared channel would disable both control paths, resulting in a complete loss of control over the device *M*.

By contrast, separating the control channels (Figure 4.5b) ensures independent operation. If one of the channels fails, the other remains functional, preserving control of the device *M*. This architectural decoupling significantly improves system reliability.

In electronics, *circuit decoupling* is commonly achieved using *decoupling capacitors*, which isolate portions of a circuit from power fluctuations caused by

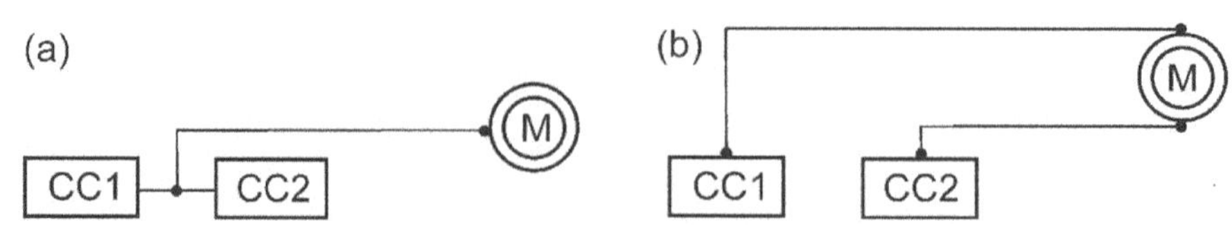

FIGURE 4.5 Improving the reliability of a control system by ensuring independent operation.

switching activities elsewhere. This prevents disturbances in one part of the system from affecting another.

Mechanical and electrical systems are also vulnerable to *vibration propagation*, which can simultaneously affect multiple components and trigger various failure modes. Mitigating the spread of vibrations through decoupling mechanisms is a well-established approach to enhancing reliability.

Furthermore, *decoupling is essential for breaking positive feedback loops* that accelerate degradation in interconnected systems (see Chapter 12). Such loops can destabilize entire networks. By interrupting these self-reinforcing interactions, decoupling helps restore and maintain system stability. As system interdependencies continue to grow, the importance of decoupling as a strategy for managing feedback loops and preserving reliability becomes increasingly critical.

4.2.9 Stochastic Separation

Guaranteeing with a specified probability the existence of separation between risk-critical factors will be termed *stochastic separation*.

Risk is often the result of the simultaneous presence of risk-critical factors. Reducing risk then depends on the existence of a separation between the risk-critical factors.

The competition of random demands for a particular resource/service on a finite time interval is a common example of risk and reliability controlled by the simultaneous presence of critical events.

Consider consumers demanding randomly a particular resource (medical resources, water vapour, particular piece of equipment, repairers), during an operation period with length L. The available resources can satisfy only a limited number of simultaneous random demands at a time. Suppose that there is only a single available resource (e.g., piece of equipment, consultant) that can satisfy only a single random demand at a time.

The random appearance of demand i is marked by s_i and the end of demand i is marked by e_i (Figure 4.6). The figure illustrates the problem for three random demands with durations d appearing at random times s_1, s_2 and s_3. The overlapping region s_3e_2 in the figure marks the simultaneous appearance of two random demands which leads to unsatisfied demand.

For complex systems (e.g., production systems, computer networks), random demands may arise from the need to repair failed components within the system. In a medical context, they can be urgent requests for specific life-saving equipment from critically injured individuals.

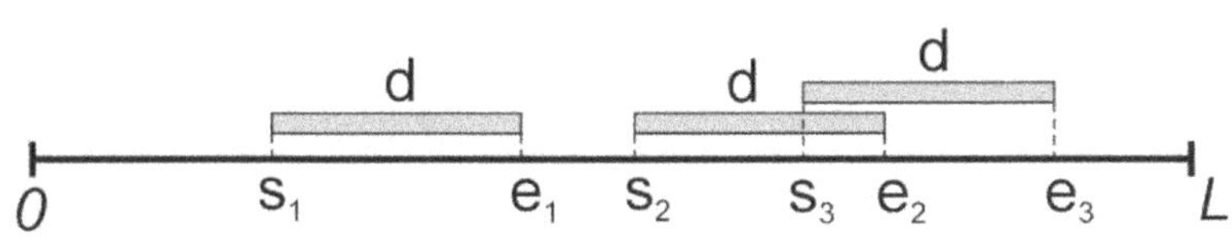

FIGURE 4.6 Random demands over a time interval.

This problem is particularly relevant in situations where, during a fixed time interval (such as a hospital shift), only a limited number of medical resources – equipment, nurses, doctors, etc. – are available to treat critically ill patients. Each resource can serve only one patient at a time. The demands for a particular type of resource occur randomly within the time, and each patient may require the resource with a certain probability.

For patients in urgent need of life-saving medical equipment, unmet demand can lead to severe consequences: deterioration of health, prolonged recovery, or even death. If two or more patient demands overlap and exceed the available quantity of a resource, some demands will go unmet. For example, if only a single consultant surgeon is available and two or more critically ill patients require immediate attention, at least one patient will face a delay. In such cases, the degree of overlap in random demands is a critical measure of risk.

Over-provisioning a particular resource type is costly and places unnecessary strain on healthcare systems, while under-provisioning increases risk beyond acceptable limits. Determining the optimal balance between the quantity of each type of medical resource and the tolerable level of risk is therefore a key issue.

Suppose that there are n consumers of a particular resource issuing random demands over the time interval (0, L) and the duration of the random demands is d. Each consumer initiates a random demand with probability $\gamma\,(0 \le \gamma \le 1)$. This means that there is a chance γ that the consumer will initiate a random demand and a chance $1-\gamma$ that there will be no random demand. It is assumed that there are m pieces of the resource servicing the random demands, which can satisfy m simultaneous random demands.

If the time fraction of the random demands is denoted by $\psi = d/L$, it can be shown rigorously (Todinov, 2018) that the expected (average) fraction of time during which exactly m random demands are simultaneously present ($m = 0,1,\ldots,n$) is given by the term

$$\frac{n(n-1)\ldots(n-m+1)}{1\times 2\times\ldots\times m}(1-\gamma\psi)^{n-m}(\gamma\psi)^m$$

which comes from the binomial expansion of the expression:

$$[(1-\gamma\psi)+\gamma\psi]^n = 1 \tag{4.1}$$

where $a = 1-\gamma\psi$ and $b = \gamma\psi$ are treated as single variables. In this binomial expansion:

$(1-\gamma\psi)^n$ is the expected fraction of time during which no random demand is present;

$n(1-\gamma\psi)^{n-1}(\gamma\psi)^1$ is the expected fraction of time during which exactly one random demand is present;

$\frac{n(n-1)}{1\times 2}(1-\gamma\psi)^{n-2}(\gamma\psi)^2$ is the expected fraction of time during which exactly two random demands are simultaneously present;

....

$(\gamma\psi)^n$ is the expected fraction of time during which exactly n random demands are simultaneously present.

The sum of all expected fractions of time is equal to one because $[(1-\gamma\psi)+\gamma\psi]^n = 1$.

Because the sum of all expected fractions is one, the total expected fraction of time during which there is unsatisfied demand ($m+1$ or more random demands are simultaneously present) is given by

$$p = 1 - \left(\begin{array}{l} (1-\gamma\psi)^n + n(1-\gamma\psi)^{n-1}(\gamma\psi)^1 + \frac{n(n-1)}{1\times 2}(1-\gamma\psi)^{n-2}(\gamma\psi)^2 \\ + \ldots + \frac{n(n-1)\ldots(n-m+1)}{1\times 2\times \ldots \times m}(1-\gamma\psi)^{n-m}(\gamma\psi)^m \end{array} \right) \tag{4.2}$$

If the users initiate their random demands with certainty: $\gamma = 1$, the model transforms into: $[(1-\psi)+\psi]^n = 1$

Stochastic separation, reducing the fraction of unsatisfied demand p, can be provided by decreasing the number n of random demands, increasing the number m of pieces of resource servicing the random demands, decreasing the time fraction $\psi = d/L$ of random demands or decreasing the probability γ of initiating a random demand.

As can be seen, the very compact model (4.1) describes the behaviour of a complex system with a number of features: (i) an arbitrary number n of consumers initiating demands, (ii) each demand randomly located along a specified time interval, (iii) each consumer initiating their random demand with a specified probability, (iv) different number of sources servicing the random demands. The model captures fully all these features.

4.2.10 Case Study: Stochastic Separation to Reduce the Expected Time for Unsatisfied Demand

Suppose that a particular manufacturing process is done by 12 machines during an 8-hour shift. Each machine stops once randomly, with probability 0.8 (80%) (at any time during the shift), for exactly 62.5 minutes, to use a particular measurement device. There is only a single measurement device available.

a. Determine the *expected time fraction of unsatisfied demand* during which there will be two or more simultaneous random demands for the measurement device.

b. What will be the expected fraction of time of unsatisfied demand if a second measurement device is introduced?

Solution:

Time fraction of a consumer demand: $\psi = \dfrac{62.5/60}{8} = 0.13$

$\gamma = 0.8;\ \gamma\psi = 0.8 \times 0.13 = 0.104;\ 1-\gamma\psi = 1-0.104 = 0.896$

For n=12 consumers, the expected time fraction of no random demands is:

$$(1-\gamma\psi)^n = (1-0.104)^{12}$$

The expected time fraction of exactly one random demand is:

$$n(1-\gamma\psi)^{n-1}(\gamma\psi) = 12 \times (1-0.104)^{11} \times 0.104$$

The expected time fraction of exactly two random demands is:

$$\frac{n(n-1)}{1 \times 2}(1-\gamma\psi)^{n-2}(\gamma\psi)^2 = \frac{12 \times 11}{1 \times 2} 0.896^{10} \times 0.104^2$$

The expected fraction of unsatisfied demand for a single available measurement device is:

$$p = 1 - \left((1-0.104)^{12} + 12 \times (1-0.104)^{11} \times 0.104^1\right) = 0.359\ (\approx 36\%)$$

The last result shows that 36% is the expected fraction of two or more simultaneous random demands. This is the time fraction during which there will be unsatisfied demand.

Stochastic separation reducing the expected fraction of unsatisfied demand can be provided by increasing the number of measurement devices.

The expected fraction of unsatisfied demand for two available measurement devices is given by:

$$p = 1 - \left((1-0.104)^{12} + 12 \times (1-0.104)^{11} \times 0.104^1 + \frac{12 \times 11}{1 \times 2}(1-0.104)^{10} \times 0.104^2\right)$$
$$= 0.12$$

Thus, the expected time fraction of three or more simultaneous random demands is 12%. This is the time fraction during which there will be unsatisfied demand if two measurement devices are introduced.

By introducing a second measurement device, stochastic separation has been made and the expected time of unsatisfied demand has been decreased three times.

4.3 SEPARATION OF FUNCTIONS, PROPERTIES, OR BEHAVIOUR

This important form of design separation can be implemented using several key mechanisms and techniques.

4.3.1 Separation of Functions

Separating functions involves assigning distinct functions to different parts of a system or component. This approach enhances reliability and reduces risk through several mechanisms.

4.3.1.1 Function Separation to Maximize Reliability

Optimizing a single component to perform multiple functions is inherently challenging. Consider a seal that serves two critical roles: (i) *Sealing*, to isolate fluids, and (ii) *Load-bearing*, to withstand tensile forces or bending moments.

Often, assigning both functions to the same part compromises reliability. For example, if a joint relies on a single seal for both fluid isolation and structural integrity, its reliability is often compromised. A common and costly design error is to combine the sealing and load-carrying functions in a single component.

To achieve high reliability, these functions must be separated. By assigning the sealing and load-bearing responsibilities to different parts, each can be optimized individually for its specific role which results in a robust and reliable joint assembly.

This principle can be illustrated through the design of a pipe or container for corrosive fluids (Figure 4.7). It is difficult to find a single material that offers both high strength and high corrosion resistance at an acceptable cost.

For instance, *low-carbon steels* provide strength at a low cost but have poor corrosion resistance while *plastics* offer excellent corrosion resistance but insufficient mechanical strength. By separating the functions, each requirement can be addressed using different materials: (1) *Steel component* provides strength, and (2) *Plastic lining* offers corrosion resistance.

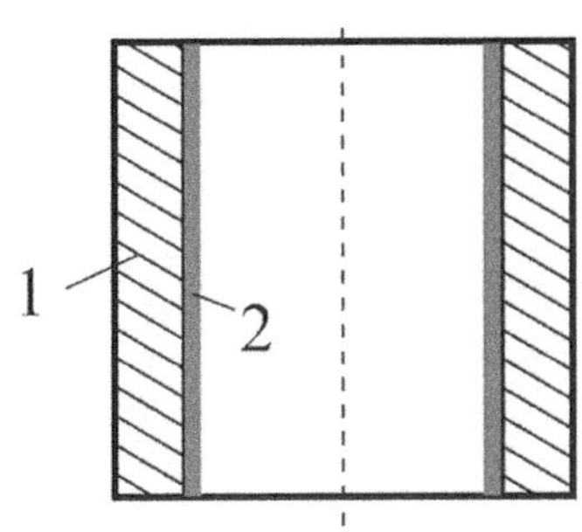

FIGURE 4.7 Separation of functions in a pipe carrying corrosive fluid.

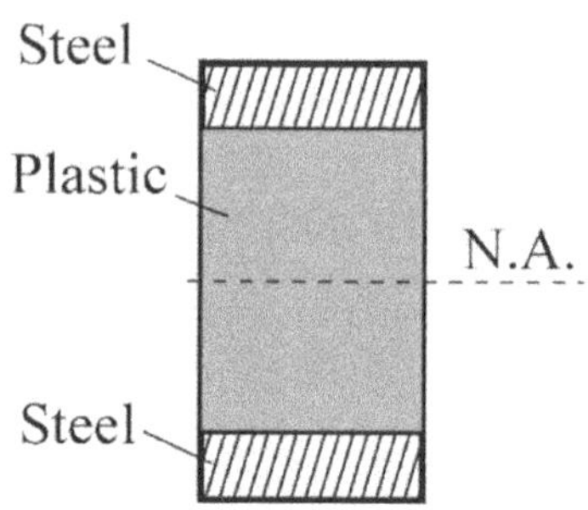

FIGURE 4.8 Separation of functions in a composite beam.

This combination achieves high reliability at lower cost which is an advantage over conventional reliability improvements (e.g., material upgrades), which typically involve a larger investment.

Further examples of this approach include plastic or resin coatings on wooden panels to protect garden sheds from rain and plastic lining of fuel tanks in road vehicles to reduce the risk of fuel leaks and fires in the event of collision.

Another illustrative example of the separation of functions principle is the design of composite beams. As shown in Figure 4.8, the composite beam consists of steel covers and a plastic core. The strong material (steel) is positioned far from the neutral axis (N.A.), where it can effectively resist most of the tensile and compressive loads. In contrast, the lightweight, low-strength plastic is placed near the neutral axis, where it contributes primarily to stability by resisting buckling. Each material serves a distinct purpose: the steel bears the structural loads, while the plastic enhances overall stability and reduces the weight of the beam.

Task specialization is a form of functional separation that involves assigning distinct, well-defined tasks to different subsystems or personnel. This division reduces complexity and confusion, thereby enhancing process robustness. For example, in quality management, separating inspection from production by assigning them to different teams ensures that monitoring and diagnosis of faults can occur independently, without interfering with operational activities.

A similar principle applies in software development, where separating functions into separate threads, processes, or containers helps contain bugs or crashes. This isolation ensures that a failure in one component does not propagate and disrupt the entire system.

4.3.1.2 Separation of Functions to Reduce Load Magnitudes

This type of functional separation reduces the overall load on individual components by distributing different load types among separate parts, rather than concentrating multiple load functions within a single component.

When a component is required to perform multiple functions simultaneously, it may be subjected to combined loading, such as axial, radial, or torsional forces, whose resultant magnitude can exceed its structural limits. The increased resultant load accelerates the rate of degradation, ultimately reducing reliability.

For example, consider a single bearing designed to carry both high-magnitude radial and axial forces. The combined load places significant stress on the bearing. Over time, this elevated loading leads to accelerated wear and reduced service life.

By separating the load-bearing functions of supporting radial and axial load, into two different bearing types, the loads are effectively decoupled: One of the bearings (e.g., a roller bearing) is dedicated to resisting *radial loads only* while the other bearing (e.g., a thrust or axial ball bearing) resists *axial loads only*.

Such a division of responsibilities reduces the load magnitude on each bearing, lowers the rate of wear, and significantly enhances the overall reliability of the bearing assembly.

Aircraft landing gear systems provide a clear example of functional separation to reduce load magnitude. Steering, braking, and load-bearing are managed by distinct assemblies, allowing each to experience lower mechanical stress and to be optimized independently. By isolating steering, braking, and shock absorption, the system gains better performance and durability.

A similar principle is applied in *seismic design*. Structural frames support vertical loads, base isolators dissipate seismic energy, and shear walls resist lateral forces. This separation of functions reduces dynamic forces transmitted to the structure, improving the overall resilience during earthquakes.

4.3.1.3 Separation of a Single Function into Multiple Components to Reduce Vulnerability to a Single Failure

This form of separation is present when a single function is distributed across multiple identical function carriers. By assigning the function to several components, the system becomes less vulnerable to a single-point failure. If a function is performed by a single component, its failure results in complete loss of the function. However, if multiple identical components each perform the same function, the failure of a single component does not compromise overall functionality.

Traditional redundancy is a widely used method for improving system reliability. This is essentially an application of the separation principle, where the same function is replicated across identical components. This type of separation effectively replaces sudden critical failure with gradual deterioration through non-critical failures.

4.3.1.4 Separation of Functions to Compensate for Deficiencies

Functional separation can also be used to compensate for the limitations of different system components. A common example is the combination of a *human operator and automation* in industrial control systems. Humans are susceptible to fatigue, delays, and errors under stress, while automation may fail in unanticipated or complex situations.

In this case, *functional separation* involves the operator and the automation system working in parallel but handling different types of decisions. The automation manages routine tasks, ensuring speed and consistency, while human operators intervene in complex or abnormal situations that require judgment or adaptation.

Another example is the combination of *batteries and supercapacitors* in energy storage systems. Batteries provide high energy density but suffer from slower response times and limited cycle life. In contrast, supercapacitors offer rapid response and long life but have low energy storage capacity.

Here, functional separation allows each component to focus on what it does best: supercapacitors handle quick load spikes, reducing stress on the batteries, while batteries supply sustained power over time. This cooperation enhances the overall efficiency and lifespan of the system.

A final example is the integration of *passive and active cooling* in electronic systems. *Active cooling* methods, such as fans and pumps, can fail due to mechanical issues or power loss. On the other hand, *passive cooling* techniques, like heat sinks and thermal radiation, are inherently reliable but less effective under high thermal loads.

Through functional separation, passive systems provide consistent baseline cooling, while active systems respond dynamically to peak loads. This arrangement improves overall thermal management and system reliability.

4.3.1.5 Separation of Functions to Prevent Unwanted Interactions

This separation mechanism, commonly referred to as *separation of concerns*, is a well-established principle in computer program design (Reade, 1989). A *concern* denotes a specific, self-contained task handled by a distinct section of the program. Functional separation in programming is typically achieved through *encapsulation*, whereby data and operations are enclosed within code modules (such as procedures or functions) that expose a clearly defined interface.

Encapsulation ensures that variables defined within a module are only accessible and modifiable within that module. This isolation prevents unintended interactions between different sections of the code, thereby avoiding side effects and hard-to-detect bugs that may occur when one part of the program alters data belonging to another.

Moreover, encapsulated modules can be updated, maintained, and tested independently, without changing other parts of the program. This modularity significantly reduces the risk of introducing errors during maintenance or extension. Effectively, each encapsulated block functions as a *black box*, with specified inputs and outputs, whose internal implementation can be developed or replaced independently, without impacting the overall program logic.

4.3.1.6 Separation of Methods to Reduce the Risk Associated with Errors in Mathematical Models

Separation is a powerful strategy for increasing the reliability of new methodologies or models. When a particular value is estimated using a complex theoretical model, its reliability can be significantly improved if the same value is independently estimated using an alternative method based on fundamentally different principles. Agreement between the results obtained from the two separate methods provides strong evidence supporting the validity of the model.

For example, computer algebra systems can be employed to validate algebraic manipulations, solve algebraic or differential equations, and perform symbolic operations such as integration, differentiation, or optimization (e.g., finding extrema of functions). Next, hand calculations on a simple, loaded component can serve as a benchmark for verifying the correct application of boundary conditions in finite element analysis software.

An illustrative example involves proving an algebraic inequality using analytical methods. The validity of the inequality can also be substantiated through Monte Carlo simulations with multiple sets of randomly generated input values for the variables involved. The agreement between the theoretical proof and the simulation results significantly reinforces confidence in the correctness of the analytical approach.

Conversely, while a Monte Carlo simulation may suggest that an algebraic inequality is correct, an analytical proof can rigorously establish that it holds for all possible combinations of the input variables. By employing multiple, independent methods to verify the same result, the likelihood of undetected modelling or computational errors is reduced, thereby improving the overall robustness and reliability of the analysis.

4.4 SEPARATION OF PROPERTIES TO COUNTER POOR PERFORMANCE CAUSED BY INHOMOGENEITY

4.4.1 Separation of Strength According to the Intensity of the Loading Stresses

Separation aimed at countering the limitations of homogeneity is essential in situations where the average property characterizing a homogeneous material is insufficient to ensure the required reliability. This approach is critical in engineering design to ensure that different regions of a component possess the strength necessary to withstand the local loading stresses.

A homogeneous material with an average strength is not optimized for spatially varying stresses and, therefore, may not provide adequate resistance throughout the entire component. For example, if a particular region of a component is subjected to high-amplitude cyclic loading, the material in that region must exhibit high fatigue resistance. In contrast, a region under static loading in a non-corrosive environment does not require enhanced fatigue resistance, since the conditions for crack initiation and propagation are absent. Because loading conditions vary across the component, the corresponding material resistance must also vary, and a homogeneous level of resistance is inadequate in such cases.

Implementing this type of separation involves deliberately introducing a non-uniform distribution of resistance properties across the component, tailored to the local type and intensity of stress experienced by each zone in the component.

Consider the two pairs of bolts in Figure 4.9, labeled '2' and '3', securing the loaded cantilever arm '1'. These bolts should not have the same strength, as bolts '2' are subjected to significant tensile stresses, while bolts '3' experience only minimal loading. Therefore, the strength of bolts '2' should be substantially greater than that of bolts '3'.

Separation of strength according to the experienced stress in a particular zone can be illustrated with objects likely to impact other objects. The likelihood that a prismatic object will come into contact with one of its corners is significant and, as a result, the object is easily damaged. If a separation of properties is applied, the corners will be made more resistant to damage. Separation of properties can be applied by reinforcing the corners with material resistant to contact damage.

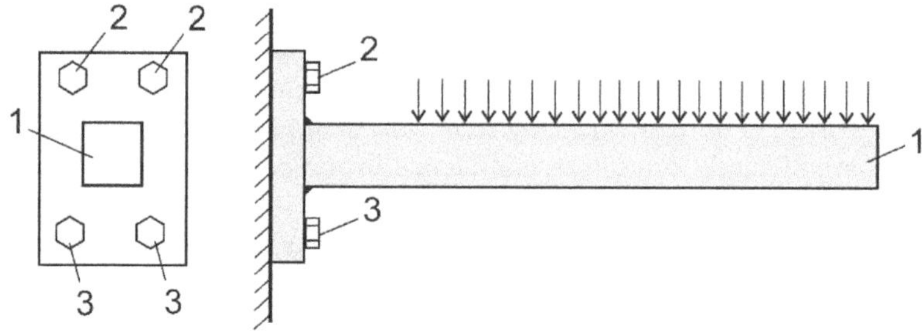

FIGURE 4.9 Loaded console arm.

4.4.2 Separation of Properties to Satisfy Conflicting Requirements

A common engineering challenge involves materials that must simultaneously exhibit conflicting properties. For example, a gear must be hard enough to withstand high contact stresses and intensive wear, yet also sufficiently soft and tough to absorb impacts. These opposing demands require different material characteristics: a hard surface to resist wear and a soft, tough core to endure shock loading. If no separation of properties is applied, a compromise must be made by selecting a homogeneous material with moderate hardness and toughness. Such a solution is inherently suboptimal, as it neither excels in wear resistance nor in impact durability. However, these conflicting requirements can be effectively satisfied by implementing a separation of properties within the component.

One practical method is *case hardening* (Kalpakjian and Schmid, 2013), where the surface layers of gear teeth are locally heated (typically by induction heating) and then rapidly quenched. This treatment produces a hard, wear-resistant surface while maintaining a tough, ductile core capable of absorbing impacts and resisting fracture. This separation of properties optimizes the gear's performance under combined loading conditions.

The demand for materials that combine such contradictory properties is growing in modern technology, leading to the development of *functionally graded materials (FGM)* (Mahmood et al., 2012). FGM achieve separation of properties through deliberate gradients in chemical composition, microstructure, and phase distribution. These gradients translate into corresponding property variations within the material.

4.5 SEPARATION BASED ON A PARAMETER, CONDITIONS, OR SCALE

Risk reduction through separation is achieved by ensuring distinct behaviour of a system or component at different values of a risk-critical parameter. By differentiating characteristics according to these parameter values, the likelihood of failure or the severity of its consequences can be significantly reduced.

One effective approach to implementing this type of separation is the introduction of deliberate weak links and stress limiters. These elements act to restrict risk-critical parameters from reaching dangerous levels. Due to its fundamental importance, this technique is discussed in greater detail, as a standalone method, in Chapter 7.

4.5.1 Separation Based on Phase Changes

Separation by exploiting phase changes is a key mechanism for controlling a common risk-critical parameter: *temperature*. Reliable separation across operating temperature ranges can be achieved by harnessing the changes in the material's physical state that occur at specific temperatures.

For example, freezing volatile and highly toxic substances during transportation significantly reduces the risk of spillage in the event of a road accident. Similarly, special alloys with low melting points serve as highly reliable triggers for fire sprinkler systems, providing a fail-safe response that is more dependable than complex triggers involving thermostats, sensors, and electronic circuits.

Phase changes of the magnetic state of materials can also be used for temperature-based separation. During induction heating for example, to prevent overheating of an object beyond a critical temperature T_{cr}, the alloy could be selected to be with Curie temperature equal to T_{cr}. At temperatures $T < T_{cr}$, the material is ferromagnetic; at temperatures $T > T_{cr}$, the material is diamagnetic. Because the magnetic properties of the material change at $T = T_{cr}$, induction heating of the material beyond $T = T_{cr}$ is not possible, therefore overheating is not possible.

Phase changes also absorb substantial amounts of heat energy and can be harnessed to provide passive thermal buffers. Thus, low-melting-point alloys can be used to absorb excess heat as they melt (latent heat), and keep the protected devices cool during short bursts of transient heat loads.

Additionally, separation based on evaporation has been applied to protect sensitive rocket control equipment from overheating. Rockets are encased in a foam shell that evaporates upon launch, providing thermal protection during critical stages.

An intriguing application of phase-change-based separation is described in U.S. Patent 4,072,159 (1978), granted to inventor Toyoki Kurosawa. The patent features an emergency valve designed to prevent the spread of fire in pipelines transporting flammable fluids.

The valve incorporates a built-in element made of thermal foamable plastic. When exposed to high temperatures, this material melts and expands into a foam that blocks the flow within the pipeline. Consequently, the section affected by fire is effectively isolated from the rest of the system, thereby halting the propagation of fire.

4.5.2 Separation of Reliability across Components and Assemblies According to Their Cost of Failure

An important risk-reduction mechanism is the allocation of reliability across components proportionally to their cost of failure. Components whose failure results in high consequences should be designed with correspondingly higher built-in reliability. This principle underlies the cost-of-failure-based design approach (Todinov, 2007). Safety-critical components, processes, and operations must achieve higher reliability than their counterparts in non-critical applications.

This domain-independent method is essential for effective risk management. Failure to apply it has led to catastrophic industrial accidents, including major oil spills, where insufficient reliability was incorporated into systems that did not correspond to the severity of their potential failure.

For example, failure of the cement used to seal a subsea oil production well can cause catastrophic environmental pollution. Therefore, the cement seal must have significantly higher reliability than cement used in less critical applications. Given the enormous cost of failure, the probability of seal failure must be minimized to reduce the potential losses proportionally. This shows why reliability separation based on cost of failure is a fundamental risk reduction strategy.

The notion that failures of components, such as cement seals, are low-probability but unpredictable events is misleading and has caused serious harm in many industries. Effective risk management requires reducing the probability of failure in proportion to the cost of failure, thereby rendering the risk acceptable. This is achieved by implementing appropriate risk controls and verifying through probabilistic assessment that the risk is sufficiently low. Unfortunately, such rigorous evaluation is often neglected for components associated with high consequences of failure.

Furthermore, separating reliability according to cost of failure plays a key role in optimizing the allocation of fixed budgets for reliability improvements. Uniformly distributing investment across all components is generally inefficient. To maximize overall risk reduction, resources should be prioritized to reduce the risk for components whose failure carries the greatest consequences and where investment yields the highest risk reduction. A specialized dynamic programming algorithm for optimal budget allocation among a given number of risk reduction options has been proposed in Todinov (2014).

4.5.3 Case Study: Separation of the Reliability of Components Based on the Cost of Failure

This principle can be illustrated with the simple control system in Figure 4.10 (Todinov, 2019c).

In the control system from Figure 4.10, there are nine sources of production (generators, compressors, pumps, wells, etc.) w1-w9, supplying a particular commodity (electricity, gas, fluid). The system consists of two identical control modules $CM1$ and $CM2$. The amount of commodity supplied per day from each production source wi is M units.

Suppose that the failure rate of control module $CM1$ is λ_{CM1} year^{-1} and the failure rate of control module $CM2$ is λ_{CM2} year^{-1}. Suppose that the two modules have the same downtime for repair of d days. For a period of operation of a years, the expected lost production due to failure of control module $CM1$ is $L_{CM1} = 8\lambda_{CM1}\, a\, d\, M$ and the expected lost production due to failure of control module $CM2$ is $L_{CM2} = \lambda_{CM2}\, a\, d\, M$. For equal failure rates of the control modules, failures of control module $CM1$ cause

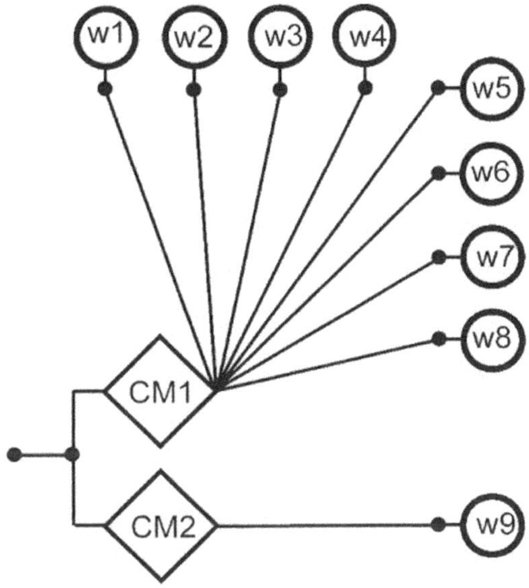

FIGURE 4.10 A production system with hierarchy based on nine production sources.

eight times greater loss of production compared to failures of control module *CM*2. The ratio of the lost production from failures over the operational time interval $(0,a)$ is

$$L_{CM1} / L_{CM2} = 8\lambda_{CM1} / \lambda_{CM2}$$

To make the losses from failures of the two control modules comparable: $L_{CM1} \approx L_{CM2}$, the failure rate λ_{CM1} of the first control module must be eight times smaller than the failure rate λ_{CM2} of the second control module: $\lambda_{CM1} = (1/8)\lambda_{CM2}$. This means that the reliability level of control module *CM*1 should be significantly higher than the reliability level of control module *CM*2.

In developing safety-critical software, separation according to the cost of failure could be implemented by increasing the amount of testing of a software module to be proportional to the cost of failure of the module.

5 Improving the Load-Carrying Capacity of Structures by Separation in Geometry

5.1 EXAMPLES OF TECHNIQUES RELATED TO INCREASING THE LOAD-CARRYING CAPACITY BY A SEPARATION IN GEOMETRY

Separation in geometry has been used for a long time in structural optimization as a foundational technique for reducing weight and enhancing the reliability and load capacity of components. It involves redistributing material and refining component shapes while adhering to specific load conditions, performance requirements, and constraints. Separation in geometry is a technique that is part of the domain-independent method of separation for improving reliability and reducing the risk of failure (Todinov, 2019c).

A common example of increasing load capacity through separation in geometry is the adaptation in plates subjected to tension, where increasing material thickness around holes mitigates stress concentration, reduces peak stresses, and improves the plate's fatigue life.

Figure 5.1 demonstrates increasing resistance to fracture by a geometric separation. Extra material is added around the hole, in the vicinity of which the magnitude of the loading stress for wide plates is approximately three times larger than the magnitude of the loading stress farther from the edge of the hole (Figure 5.1a). Adding extra material results in significantly reduced stresses at the edge of the hole and improved fatigue resistance (Figure 5.1b). This method has application for reinforcement of punched plastic pockets (Figure 5.1c).

Another common example of increasing the load capacity of components through separation in geometry is the tapered profile of a cantilevered beam, where thickness decreases from the fixed end toward the free end.

5.1.1 Case Study: Separation in Geometry for a Cantilever Beam

Consider the cantilever beam with length L and uniform rectangular cross section with thickness t and width b (Figure 5.2a) which has been loaded by a force with magnitude P. Suppose that the tensile strength of the material is σ_s. According to the

DOI: 10.1201/9781003679301-5

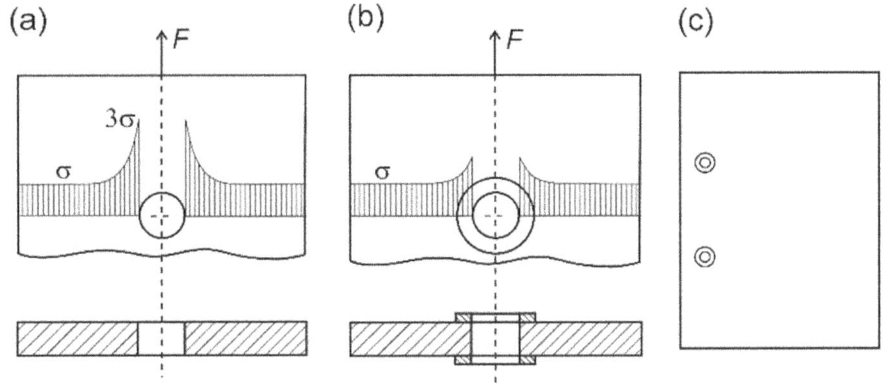

FIGURE 5.1 The separation in geometry reduces the maximum stress in the vicinity of a hole.

simple beam theory (Gere and Timoshenko, 1999), the maximum tensile stress σ_t acting on the beam, is at the cantilever support and is given by

$$\sigma_t = \frac{6PL}{bt^2} \tag{5.1}$$

Consequently, the maximum magnitude of the loading force which the beam 'a' (in Figure 5.2a) can sustain is

$$P_{\max,a} = \frac{\sigma_s bt^2}{6L} \tag{5.2}$$

Suppose that the beam has been tapered in the way shown in Figure 5.2b, such that the right end of the beam has thickness $0.5t$ and the left end has thickness $1.5t$. If the width b and the length L of the initial beam remain unchanged, the volume of material $V_b = \frac{(0.5t + 1.5t)}{2} L \times b = tLb$ used for the beam in Figure 5.2b is equal to the volume of material $V_a = tLb$ used for the beam in Figure 5.2a. However, the maximum stress the beam in Figure 5.2b can sustain is now 2.25 times greater than the maximum stress characterizing the beam in Figure 5.2a:

$$P_{\max,b} = \frac{\sigma_s b(1.5t)^2}{6L} = 2.25\, P_{\max,a} \tag{5.3}$$

The load-carrying capacity of the beam has been increased, and the risk of failure has been reduced by a separation in geometry, at the same volume of material.

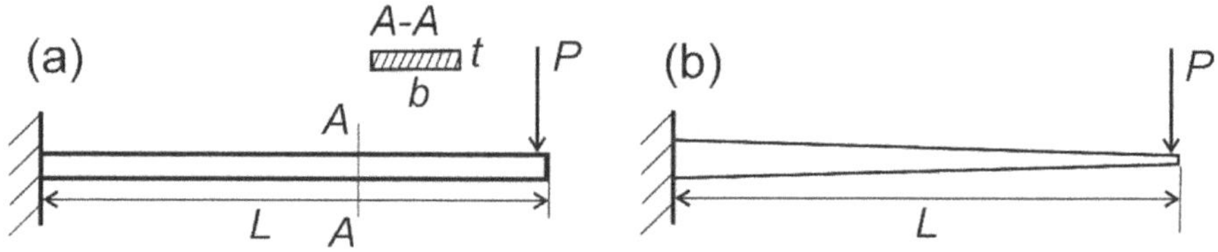

FIGURE 5.2 Separation in geometry for improving the reliability of a cantilever beam.

The beam is made thicker towards the cantilevered end where the stresses are the largest and thinner towards the free end where the stresses are the smallest. With the advancement of technology, the importance of separation in geometry is increasing. By using 3D-printing technique, for example, components with complex shapes can now be produced using less material.

Topology optimization (Bendsøe and Sigmund, 2003), developed as a mainstream design optimization technique for achieving lightweight structures, can also be regarded as a special case of the domain-independent approach 'separation in geometry'. In recent years, topology optimization has been widely applied to generate lightweight designs (Zhu et al., 2021; Walton and Moztarzadeh, 2017; Zhao et al., 2024). Several methods have been developed, including the evolutionary structural optimization method, the density method, and the level set method (Zhu et al., 2021).

Advancements in manufacturing technologies, such as additive manufacturing (3D printing) (Bikas et al., 2016; Yang et al., 2016; Mandolini et al., 2022) and advanced metal forming (Rosenthal et al., 2020) helped produce component shapes difficult to create using other methods. It also allows the integration of multiple functions into a single component, further improving load capacity at the same or reduced weight.

Aggregation of structural components made from different materials has also been exploited to obtain composite beams with superior flexural strength, exceeding the strength of the individual components. Examples include glass fibre and carbon fibre composite beams, beams made of steel-reinforced concrete, and sandwich beams. Sandwich beams, for example, utilize the principle of separation of functions in design: strong material located away from the neutral axis resists most of the load, while light and low-strength material near the neutral axis provides stability (Buragohain, 2017; Beer et al., 2002; Gere and Timoshenko, 1999; Budynas, 1999).

Aggregation techniques have been exploited in some structural designs where steel plates are aggregated by welding to form reinforced beams (Zhou et al., 2021). This increases the overall load-carrying capacity of the beam by allowing it to handle higher bending moments and shear forces. The structural behaviour of aggregated steel box beams filled with concrete has been investigated in Su et al. (2025).

In this chapter, we will introduce a different type of aggregation: the aggregation of identical structural components made of the same material into fewer identical structural components with larger cross-sections.

The method of aggregation presented in this chapter works for beams with arbitrary cross-sectional shapes, and the reductions in maximum stress and deflection are impressive. For loaded components with arbitrarily shaped cross-sections, the method of aggregation has been recently introduced in (Todinov, 2025c). Multiple loaded elements are widely used in engineering and construction but in stress analysis literature, there is a lack of analysis regarding the effect of aggregation. This is a striking omission considering that aggregation is a relatively simple idea.

Discussion related to the aggregation idea is also absent in structural engineering textbooks and papers related to the optimization of loaded beams. This critical discussion is also missing in the structural reliability literature.

The lack of discourse on the effects aggregation of components with arbitrary cross-sections constitutes a significant knowledge gap, highlighting the need for further exploration.

We need to point out that although the proposed aggregation method is a kind of separation in geometry, it is a fundamentally different approach from the standard topology optimization approach and is developed without the help of any topology optimization algorithms. Topology optimization entails significant computational overhead due to the large number of iterative simulations required. In contrast, the proposed aggregation method does not involve any iterative simulations and incurs minimal computational overhead. Unlike topology optimization, the proposed method of aggregation is exact and guarantees attainment of the optimal solution. Considering also the drastic weight saving achieved by this method, it is a valuable optimization method, part of the general domain-independent approach 'separation in geometry'.

Bending, or flexural loading, is a very common type of loading. Elements with increased load capacity in bending are less likely to experience fatigue or premature failure due to overloading or due to sudden application of loads. This results in a longer service life for the structure, reduced maintenance costs, and enhanced safety.

5.2 Method of Aggregation: Basic Concepts of the Aggregation Method

Consider an affine transformation with different scale factors for each axis. This transformation applies separate scaling values to the x- and y-coordinates of a cross-section S (Figure 5.3), stretching or compressing it non-uniformly and changing its aspect ratio. The affine transformation with different scale factors is different from a transformation with uniform scaling which applies a single scale factor for all axes.

As a result of the affine transformation with scaling factors $k_x \geq 1$ and $k_y \geq 1$, the cross-section S' is obtained from the cross-section S (Figure 5.3).

The matrix multiplication

$$\begin{bmatrix} x_2 \\ y_2 \end{bmatrix} = \begin{bmatrix} k_x & 0 \\ 0 & k_y \end{bmatrix} \begin{bmatrix} x_1 \\ y_1 \end{bmatrix} = \begin{bmatrix} k_x x_1 \\ k_y y_1 \end{bmatrix} \tag{5.4}$$

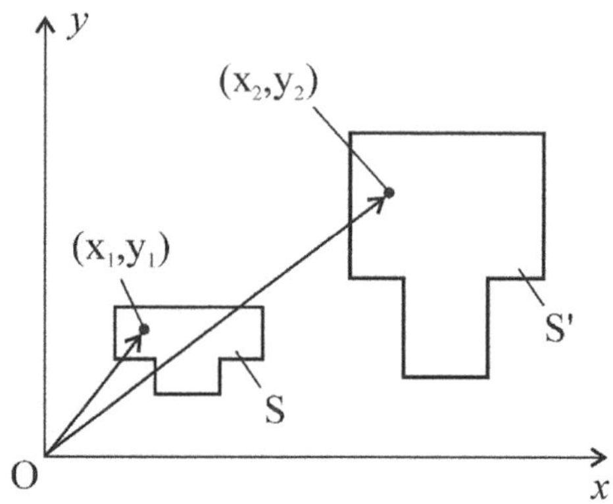

FIGURE 5.3 Affine transformation with different scale factors along the coordinate axes.

describes the transformation of any vector $\begin{bmatrix} x_1 \\ y_1 \end{bmatrix}$ from the original cross-section S into the vector $\begin{bmatrix} x_2 \\ y_2 \end{bmatrix}$ from the resultant cross-section S'. The matrix $\begin{bmatrix} k_x & 0 \\ 0 & k_y \end{bmatrix}$ is the scaling matrix defining the affine transformation. Because k_x and k_y are not necessarily equal, the original cross-section S and final cross-section S' are not necessarily geometrically similar. They will be geometrically similar only if $k_x = k_y$.

Consider now a non-aggregated structure (Figure 5.4a) composed of n_1 cantilever beams and an aggregated structure (Figure 5.4b) composed of a smaller number n_2 of cantilever beams ($n_2 < n_1$) with larger cross sections.

Let the non-aggregated structure be subjected to a total load P and the load per beam in the non-aggregated structure be equal to P / n_1 (Figure 5.4a). Similarly, let the aggregated structure be also subjected to a total load P and the load per beam in the aggregated structure be equal to P / n_2 (Figure 5.4b). The key assumptions in deriving the next theoretical results are as follows:

- The cross-sectional geometries of the beams, both in the non-aggregated and aggregated configurations, are of arbitrary shape. The cross-section of the beams from the aggregated structure is derived through an affine transformation using scaling factors $k_x \geq 1$ and $k_y \geq 1$ along the width and the depth of the beam, as illustrated in Figures 5.4c and 5.4d.
- To simplify the theoretical formulation, it is assumed that the cross-section remains uniform along the entire length of the beam.
- The material of the beam is homogeneous.
- A uniform distribution of load is assumed across the individual load-bearing elements.

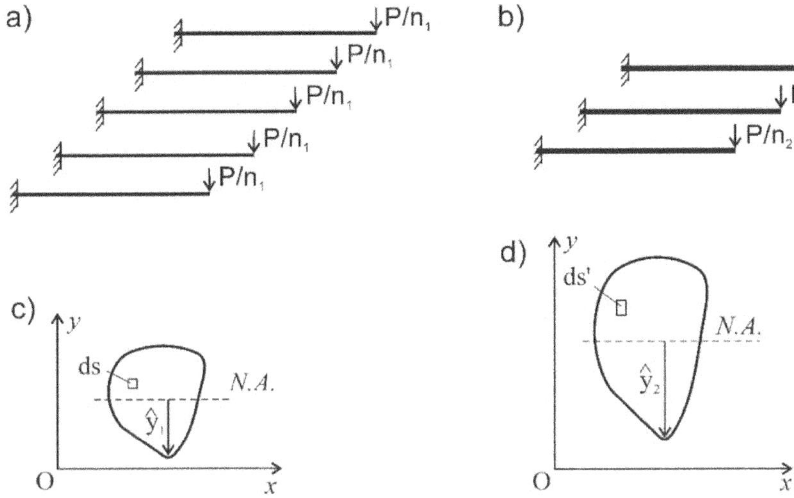

FIGURE 5.4 (a) A non-aggregated structure built on n_1 identical cantilever beams; (b) an aggregated structure built on n_2 identical cantilever beams ($n_2 \leq n_1$); (c) a beam cross section from the non-aggregated structure; (d) a beam cross-section from the aggregated structure.

These assumptions are introduced to simplify the theory but do not necessarily limit the application of the aggregation method. As a general design approach, the method remains valid even in cases where some of these assumptions are relaxed. Specifically, the method still works for structures with non-uniform cross-sections along the beam length or in cases where the load distribution is non-uniform across the load-bearing elements.

5.3 IMPROVING THE LOAD-BEARING CAPACITY AT A CONSTANT VOLUME OF MATERIAL AND THE SAME TOTAL LOAD

Suppose that the non-aggregated structure in Figure 5.5a includes n_1 identical beams with arbitrary shape of the cross-section while the aggregated structure in Figure 5.5b includes fewer number n_2 of identical beams ($n_2 < n_1$).

The beams from the non-aggregated and aggregated structure have cross- sections with coefficients of proportionality $k_x \geq 1$ and $k_y \geq 1$ between the linear parameters (see the example in Figures 5.4c and 5.4d). The coefficients of proportionality k_x, k_y, have now been selected such that the total volume of material in the non-aggregated structure is equal to the total volume of material in the aggregated structure.

Let $\sigma_{t,1}, \sigma_{t,2}$ denote the maximum tensile stress in the non-aggregated and aggregated structure correspondingly, while δ_1, δ_2 denote the maximum deflections in the non-aggregated and aggregated structure. The length of the elements from the non-aggregated and aggregated structure is equal to L. The total load for both the non-aggregated and aggregated structure is the same.

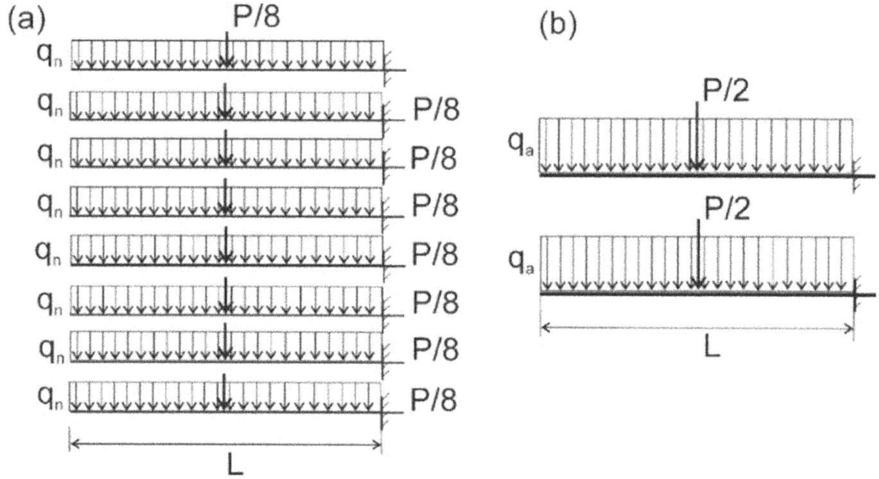

FIGURE 5.5 Testing the aggregation method on statically determinate cantilevered I-beams.

The following relationships exist between the maximum tensile stresses $\sigma_{t,1}, \sigma_{t,2}$ and deflections δ_1, δ_2 of a non-aggregated structure built on n_1 identical beams and an aggregated structure built on n_2 identical beams ($n_2 < n_1$), (the total volume of material and the total load are the same):

$$\sigma_{t,1}/\sigma_{t,2} = k_y \tag{5.5}$$

$$\delta_1/\delta_2 = k_y^2 \tag{5.6}$$

The rigorous proof of these assertions, which are new results in stress analysis, has been provided in Todinov (2025c). Aggregation under the same total volume of material results in a smaller maximum tensile stress ($\sigma_{t,2} < \sigma_{t,1}$) and smaller maximum deflection ($\delta_2 < \delta_1$). Another advantage of the aggregated structure over the non-aggregated one is the smaller number of connection points (anchors) during assembly. The same analysis and the same results (5.5) and (5.6) are also obtained for simply supported beams.

Assertions (5.5) and (5.6) mean that the reduction in the maximum stress and deflection depends on the scaling factor k_y along the depth of the beam and is independent of the scaling factor k_x along the width of the beam. This means that if the scaling factor along the depth of the beam is $k_y = 1$, then $\sigma_{t,1}/\sigma_{t,2} = 1$ and $\delta_1 / \delta_2 = 1$, and although $k_x > 1$, the aggregation into fewer load-carrying elements with larger cross-sections *will not entail any reduction of the maximum stress and deflection.*

The results (5.5) and (5.6) cannot be directly inferred from the well-known principle in solid mechanics that increasing a beam's cross-sectional depth increases the second moment of area thereby reducing stress and deflection. This is because, as the depth of the cross-section increases, the load carried by the beam also increases.

Indeed, during aggregation, the total load is distributed among a smaller number of beams. Thus, the effects from increased beam depth and increased load counteract one another, and determining which effect dominates requires extensive analytical treatment.

The results (5.5) and (5.6) also cannot be directly inferred from the reasoning that increasing the total second moment of area of the aggregated structure, under the same total load and total volume of material, automatically leads to a smaller maximum stress. It can be shown that an aggregated structure with a larger total second moment of area, under the same total load and with the same total volume, can exhibit a larger maximum stress compared to the non-aggregated structure. This occurs, for example, when the beams in the aggregated structure have different shapes, yet each beam has the same volume and carries the same share of the total load.

We need to point out that for the sake of simplicity, a uniform cross-section along the beam's length is intentionally chosen to simplify the theoretical exposition of the aggregation method. This choice in no way diminishes the advantages of the proposed method.

Aggregation must be applied judiciously, considering structural redundancy and operational conditions. For example, if a structure with four supporting elements must remain functional after the failure of any element, a 4-to-2 aggregation may be unsuitable. However, by no means does aggregation promote structural collapse. In fact, an aggregated structure exhibits significantly increased load-bearing capacity and a much larger margin of safety than its non-aggregated counterpart.

The proposed aggregation method can find important applications for increasing the load-bearing capacity and reducing the volume of material of portal frames and supporting balcony consoles. A limitation of the method of aggregation is that it can be applied only for elements loaded in bending.

5.4 Testing the Aggregation Method Using Finite Element Analysis

The method of aggregation was tested on structures composed of statically determinate cantilevered structures and statically indeterminate Π -frames. The verification of the aggregation method was conducted using steel components made of high carbon steel, heat treated to create a strong elastic material with Young's modulus of 200 GPa and a Poisson's ratio of 0.3. The steel components have been subjected to 8-to-2 aggregation. The maximum von Mises stress and the maximum deflections have been analysed using the finite element analysis software Abaqus/CAE 2021.

5.4.1 Testing the Aggregation Method on Statically Determinate Cantilevered Beams

One of the ends of the statically determinate cantilevered beam is fixed, while the other end is free (Figures 5.5a and 5.5b). The distance between the fixed support and the end of the beams is $L = 100$ mm for both the non-aggregated (Figure 5.5a) and aggregated beams (Figure 5.5b).

The cross-sections of the non-aggregated and aggregated beams are according to Figures 5.6a and 5.6b, respectively. The dimensions of the cross-sections have been selected such that the total volume of material in eight non-aggregated beams (Figure 5.5a) is exactly equal to the total volume of material in two aggregated beams (Figure 5.5b). For this reason, the scaling factors for all cross-sections were $k_x = k_y = 2$.

The same total load *P*=360N has been applied to both the aggregated and non-aggregated structure. As a result, each beam from the non-aggregated structure in Figure 5.5a (8 beams) is loaded with a force of 360/8=45N, distributed uniformly along the top surface of the beam. Each beam of the aggregated structure (2 beams; Figure 5.5b) is loaded with a force of 360/2=180N (8x45N = 2x180N=360N), distributed uniformly along the top surface of the beam. The global size of the selected hexahedral mesh was 0.0001 mm for the non-aggregated beams and 0.0002 mm for the aggregated beams.

Table 5.1 lists the maximum von Mises stress and the maximum deflection characterizing non-aggregated and aggregated cantilever structures. From the table, it can be concluded that for the same total applied load (360 N), the aggregated structure experiences significantly lower maximum von Mises stress and maximum deflection compared to the non-aggregated structure. The ratio of the maximum von Mises stresses of the non-aggregated and aggregated structures is 350/179 = 1.955, which is close to the scaling factor $k_y = 2$ along the depth of the beam. The ratio of the deflections of the non-aggregated and aggregated structures is 1.92/0.48 = 4, which is equal to the square of the scaling factor $k_y^2 = 4$ along the depth of the beam. These values correlate well with the theoretical predictions regarding the maximum tensile stress and maximum deflection for cantilever beams. The deviation of the ratio of stresses from $k_y = 2$ is present because von Mises stresses were calculated in the finite element analysis rather than the maximum tensile stress. Unlike the maximum tensile stresses, the von Mises stresses also capture the impact of the shear stresses and this is the reason for the deviation from the theoretical results which are related to the maximum tensile stress only.

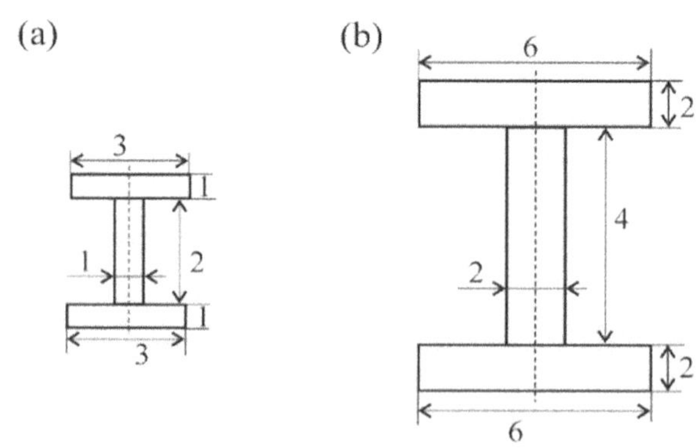

FIGURE 5.6 Cross sections of the (a) non-aggregated and (b) aggregated I-beams. All dimensions are in millimetres.

TABLE 5.1
Finite element analysis results for the maximum von Mises stress and deflection of non-aggregated and aggregated cantilever beams

Calculated quantity	Non-aggregated structure	Aggregated structure
Maximum von Mises stress	350 MPa	179 MPa
Maximum deflection	1.92 mm	0.48 mm

5.4.2 Applications of Portal Frames

Portal frames are one of the most widely used structural systems due to their efficiency, cost-effectiveness, and adaptability. They provide large clear spans without the need for internal columns, making them economical in material consumption compared to alternative systems. In addition, portal frames allow rapid fabrication and erection, offer flexibility in cladding and roofing design, and resist lateral forces effectively through their rigid connections.

Because of these qualities, portal frames are the preferred choice for many building types, including warehouses, aircraft hangars, industrial facilities, agricultural buildings, sports halls, retail outlets, transport depots, exhibition spaces, community centres, and cold storage facilities.

Here are the reasons why portal frames are ideal for these applications:

- Warehouses require large unobstructed floor areas for storage and forklift movement. Portal frames deliver clear spans of 50 m or more without internal columns.
- Aircraft hangars demand very wide and tall enclosures to accommodate aircraft. Portal frames provide economical long-span structures compared with trusses or space frames.
- Industrial buildings need flexible layouts for machinery and production lines. The open floor space created by portal frames makes reconfiguration easy.
- Agricultural buildings such as barns, sheds, and grain stores, often require quick, low-cost construction in remote areas. Portal frames are durable, easy to assemble, and resistant to wind and snow loads.
- Sports halls require wide, column-free playing surfaces. Portal frames provide high internal clearance and unobstructed sightlines for players and spectators.
- Retail stores and supermarkets depend on open layouts for shelving flexibility and smooth customer flow. Portal frames allow large spans at low cost for commercial developers.
- Bus and train depots need wide bays for vehicle entry, exit, and servicing. Portal frames accommodate large doors and roof heights without internal obstructions.
- Exhibition halls require flexible layouts with temporary partitions. Portal frames support open-span roofs, making spaces adaptable for varied events.

- Cold storage facilities demand airtight enclosures with minimal thermal bridging. Fewer columns and joints in portal frames simplify insulation and reduce heat loss.
- Community halls and event centres must serve multiple purposes (seating, staging, partitions). Portal frames offer such versatility.

5.4.3 Testing the Aggregation Method on Centrally Loaded Portal Frames

Testing of the aggregation method was conducted on centrally loaded and side- loaded scaled-down portal frames (Π-frames), with a simplified geometry. The aggregation selected for the Π-frames was 8 to 2. Eight centrally loaded frames (Figure5.7a) with a square cross-section with side 1mm are aggregated into two centrally loaded portal frames (Figure 5.7b) with a square cross-section with side 2 mm. Figure 5.7 also shows the dimensions of the Π-frames in mm. Because the total length of the frames is the same, the same total volume of material characterizes both structures. Both structures are also subjected to the same total loading force of $P = 160N$. This means that the load applied on an individual non-aggregated Π-frame is 160/8=20N, uniformly distributed over the top surface of the non-aggregated frames (Figure 5.7a).

The load applied on an individual aggregated Π-frame is 160/2=80N, uniformly distributed over the top section of the aggregated frames in Figure 5.7b. The Young's modulus of the material for all Π-frames is 200 GPa and the Poisson's ratio is 0.3.

The maximum von Mises stresses and deflections of the Π-frames were determined by using finite element analysis in Abaqus/CAE 2021. The global size of the selected hexahedral mesh was 0.0001mm for the non-aggregated frames and 0.0002mm for the aggregated frames.

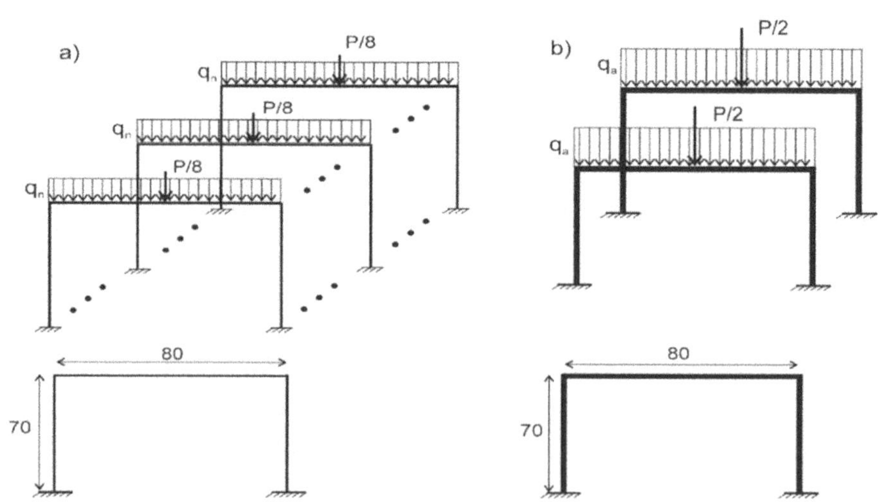

FIGURE 5.7 Eight-to-two aggregation of centrally loaded portal frames (Π -frames). All dimensions are in millimetres.

TABLE 5.2
Results for the maximum von Mises stress and maximum deflection for centrally loaded non-aggregated and aggregated Π-frames

Calculated quantity	Non-aggregated structure	Aggregated structure
Maximum von Mises stress	570.7 MPa	302.3 MPa
Maximum deflection	3.39 mm	0.8 mm

Table 5.2 lists the maximum von Mises stresses and deflections. The analysis of the results in Table 5.2, for centrally loaded Π-frames, shows that aggregated frames experience significantly smaller maximum von Mises stresses and deflections compared to non-aggregated frames.

The ratio of the maximum von Mises stresses of the non-aggregated and aggregated structures is 570.7/302.3 = 1.89 while the ratio of the maximum deflections of the non-aggregated and aggregated structures is 3.39/0.8 = 4.24.

5.4.4 Testing the Aggregation Method on Side-Loaded Portal Frames

Testing of the aggregation method was also conducted on a side-loaded, scaled-down Π-frames. The aggregation selected for the side-loaded Π-frames was also 8 to 2. Eight loaded frames (Figure 5.8a) with square cross-sections with side 1 mm are aggregated into two side-loaded frames (Figure 5.8b) with a square cross-section with side 2 mm. Figure 5.8 gives the dimensions of the side-loaded Π-frames (in mm). Both structures were subjected to the same total loading force of $P = 56N$. The load applied on an individual non-aggregated Π-frame is 56/8=7N, uniformly distributed over the right vertical section of the non-aggregated frame (Figure 5.8a).

The load applied on an individual aggregated Π-frame is 56/2=28N, uniformly distributed over the right vertical section of the aggregated frame (Figure 5.8b). The Young's modulus of the material for all side-loaded Π-frames is 200 GPa and the Poisson's ratio is 0.3.

Using finite element analysis in Abaqus/CAE 2021, the maximum von Mises stresses and deflections of the side-loaded Π-frames were determined. The global size of the hexahedral mesh was 0.0001 mm for non-aggregated frames and 0.0002 mm for aggregated frames.

Table 5.3 lists the results related to the maximum von Mises stresses and deflections. Their analysis shows that the aggregated frames also experience significantly smaller von Mises stresses and deflections compared to non-aggregated frames.

Thus, the ratio of the maximum von Mises stresses of the non-aggregated and aggregated frames is 634.5/315.4 = 2.01 while the ratio of the deflections of the non-aggregated and aggregated structures is 3.94/0.97 = 4.06.

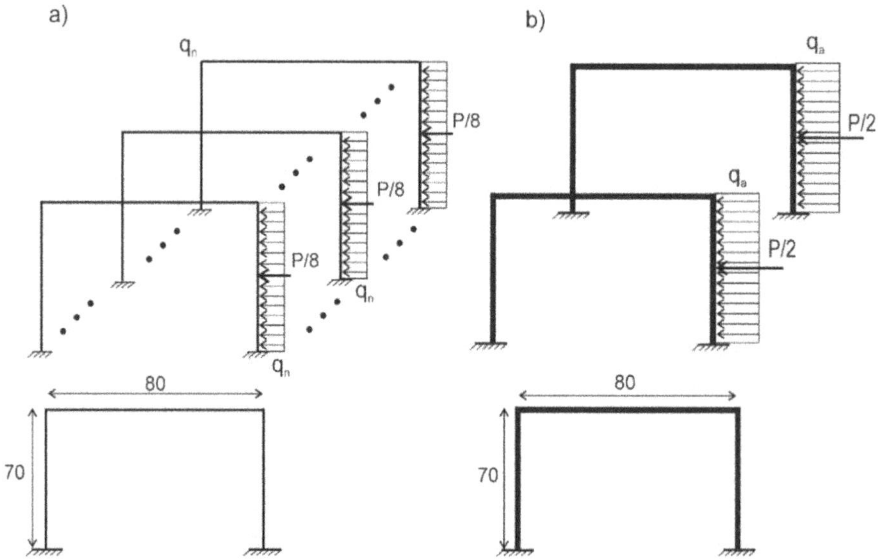

FIGURE 5.8 Eight-to-two aggregation of side-loaded portal frames (Π-frames). All dimensions are in millimetres.

TABLE 5.3
Results for the maximum von Mises stress and maximum deflection for side-loaded non-aggregated and aggregated Π-frames

Calculated quantity	Non-aggregated structure	Aggregated structure
Maximum von Mises stress	634.5 MPa	315.4 MPa
Maximum deflection	3.94 mm	0.97 mm

These simulation experiments also confirm that aggregating statically determinate cantilever beams and statically indeterminate portal frames leads to a drastic increase in the load-carrying capacity of the structures, as measured by both maximum von Mises stress and maximum deflection.

6 Improving Reliability and Reducing Risk by the Method of Segmentation

6.1 CLASSIFICATION OF SEGMENTATION TECHNIQUES: SEGMENTATION AS A PROBLEM-SOLVING STRATEGY

The underlying idea of the method of segmentation is *to prevent failure modes and reduce the vulnerability to a single failure, by dividing an entity into smaller parts.* A classification of techniques for improving reliability and reducing risk by segmentation has been presented in Figure 6.1 (Todinov, 2017b).

Segmenting and diversifying an investment portfolio by including a wide range of non-correlated stocks is a well-established strategy for mitigating financial risk. As the number of non-correlated assets increases, portfolio variance, or volatility, which is a common measure of financial risk, decreases significantly (Teall and Hasan, 2002).

A similar concept is applied in cybersecurity through *micro-segmentation*, which involves isolating different applications and components of a network to enhance security. This approach has been explored by Mämmelä et al. (2016).

Segmentation is also one of the foundational principles in the TRIZ methodology for inventive problem-solving. However, in TRIZ, segmentation is primarily used to resolve technical or physical contradictions and generate creative solutions, not explicitly for improving reliability or reducing risk. While some examples cited in TRIZ literature do apply segmentation for enhancing reliability, TRIZ literature lacks detailed discussion on the underlying techniques and mechanisms through which segmentation actually reduces risk.

Segmentation should not be understood in the rather narrow context as size segmentation or time segmentation. Segmentation extends beyond the mere physical division of an object. It can also involve the introduction of distinct boundaries with different properties within an otherwise homogeneous component, even when no physical separation is present.

A practical example is the use of stiffening rings welded at regular intervals around an underwater pipeline. These rings serve to localize and contain potential collapse between two adjacent rings, thus minimizing the extent of structural damage. In this case, segmentation is achieved not through physical separation of the entire structure, but by introducing discrete boundary elements with specific functional properties.

DOI: 10.1201/9781003679301-6

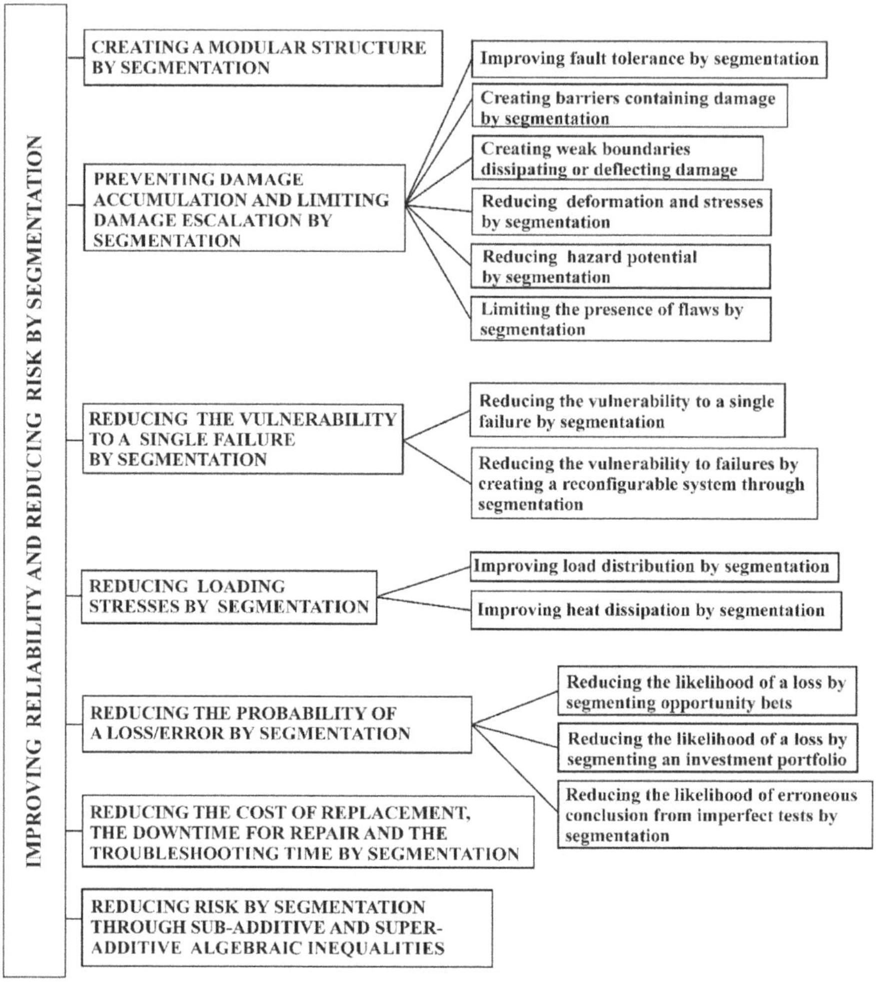

FIGURE 6.1 Various techniques for improving reliability and reducing risk by segmentation.

Another form of segmentation is the *logical segmentation*. This is present when a system is divided conceptually, rather than physically. A computer network segmented into logical subnets illustrates this well. While no physical reduction in size or complexity is present, logical segmentation introduces new boundaries that may, in fact, increase overall system complexity.

Segmentation as a problem-solving tool predates TRIZ and has been recognized for a long time in other domains. One of the most powerful strategies leveraging segmentation is the *divide-and-conquer* approach. This strategy breaks a complex problem into smaller, more manageable sub-problems that can be solved independently. The solutions to these sub-problems are then combined to resolve the original, larger problem. This approach underpins various computational algorithms, such as the *heapsort* algorithm, which guarantees a worst-case time complexity of $O(n \log_2(n))$, unlike quicksort, which can degrade to $O(n^2)$ in the worst case (Sedgwick, 1992).

Segmentation is also central to methods in systems engineering. In system reliability analysis, for example, the *decomposition method* (Todinov, 2016) divides a complex system into four simpler subsystems based on the distinct operational states of two key components ('working', 'failed'). As a result, four subsystems are obtained from the decomposition. The overall reliability is then determined by aggregating the reliability of the four subsystems.

In project management, segmentation also plays a foundational role. Breaking down a complex task into smaller, more manageable sub-tasks simplifies execution, improves control, and reduces the risk of error.

The same principle is used widely in mathematics: evaluating a complex expression becomes less error-prone when it is broken down into simpler terms that are computed separately. In mathematics, segmentation also lies at the core of integration. By dividing a complex domain into multiple simpler sub-domains, one can calculate area or volume easily.

In all these domains, segmentation, whether physical, logical, or conceptual, proves to be an enduring and versatile problem-solving strategy.

Segmentation techniques can also be applied to improve reliability and reduce risk. The idea that segmentation can serve as a method for achieving reliability improvement and risk reduction has been proposed in Todinov (2015) and further developed in Todinov (2017b).

6.2 CREATING A MODULAR SYSTEM BY SEGMENTATION

A key mechanism by which segmentation enhances reliability and reduces risk is through the creation of self-contained modules that are internally cohesive. These modules interact with one another via clearly defined input and output interfaces. One major advantage of such a modular system is fault isolation: errors or failures are confined within individual modules, making them easier to identify, diagnose, and repair compared to systems without segmentation. Faulty modules can be replaced or bypassed without the need to overhaul the entire system.

Modularization through segmentation is a powerful strategy for managing complexity in modern software systems, offering improved transparency and reliability. Segmentation also forms the foundation of *nestedness*: a hierarchical structure in which modules are composed of smaller modules, which in turn are made up of even smaller modules, and so on. This nested organization enables the construction and management of highly complex systems involving a large number of components. For example, a network can be defined as a collection of nodes and their interconnections (links), where each node itself represents another network of sub-nodes and links, and so forth.

Nestedness also supports *economy of thought*, allowing complex systems or algorithms to be described and understood more easily, which in turn reduces the likelihood of errors. This is particularly evident in recursive algorithms, which leverage nestedness to break down a problem of size n into two sub-problems of smaller size, for example, of size $n/2$. These are then recursively subdivided into smaller parts of size *n/4*, *n/8*, and so on, until trivial sub-problems are reached, each of which can be solved easily.

A modular structure enables efficient coordination among developers working across different geographic locations, allowing the assembly of complex software systems in a shorter time frame and with fewer errors. Modular development allows multiple developers to work in parallel, facilitating task specialization. For instance, expert developers can focus on user interface design, numerical routines, or database integration, depending on their strengths.

In software development, modularization shifts the focus from the internal implementation of each module to the interaction between modules. Each module can be treated as a 'black box' with well-defined inputs and outputs. This abstraction simplifies algorithm design and facilitates the selection of appropriate methods without requiring detailed knowledge of each module's internal workings.

Modularity also supports the reuse of standardized components, whose reliability has already been established, instead of developing them from scratch. This is particularly beneficial in building reliable software. Many high-quality, rigorously tested libraries are available for tasks such as creating user interfaces, solving algebraic and differential equations, or finding function extrema. These components, developed by expert teams, typically require only a specific input format and can be seamlessly integrated. Using comprehensively tested components significantly reduces development time and minimizes the likelihood of introducing software bugs.

In contrast, monolithic code tends to be highly interconnected, difficult to read, and prone to errors. Such code often devolves into 'spaghetti code', where even minor changes can trigger hard-to-isolate bugs throughout the software system.

Modular code is inherently easier to test and maintain. Individual modules typically consist of a small number of statements, which simplifies debugging and dramatically reduces the incidence of programming errors. Modules can be tested independently, even in early development stages, by using placeholder return values. Once the module's logic is implemented, it can be rigorously tested in isolation. If errors occur, they can be quickly traced to a specific module, enabling targeted debugging. In contrast, debugging thousands of interconnected lines of code is a much more challenging and error-prone process.

Finally, modular software architecture allows individual functions to be easily updated or replaced with improved or more efficient versions without disrupting the overall system structure or its established interconnections. This flexibility supports ongoing optimization and long-term maintainability.

A similar principle applies in mechanical systems, where *the operational lifespan of a component can be extended by segmenting it into replaceable sections*. This approach allows for the selective replacement of worn or failed sections without the need to replace the entire component. The benefits are significant when high-wear zones are identified and segmented accordingly. For example, in conveyor systems used to transport tiles for polishing, continuous operation leads to significant surface wear due to abrasive contact. By designing the conveyor belt with replaceable surface segments, specifically in the zones under the most stress, the need to replace the entire belt is eliminated. Instead, only the worn segments are repaired or replaced, such as restoring the rubber coating on steel plates. Similar strategies are widely used in engineering, such as replaceable inserts in journal bearings or pads in friction brakes.

Segmenting a complex mechanical device into distinct functional blocks also simplifies maintenance and significantly reduces downtime. Faults can be quickly localized, and failed modules replaced without dismantling the entire system. This modularity improves both availability and efficiency. Thus, a segmented battery pack for an electric vehicle allows isolation of a fault and continued safe operation without the need to replace the entire battery pack. Segmentation of the battery pack also provides thermal separation and limiting the spread of damage from faulty parts.

Segmentation also enhances information management. By logically dividing critical information into categories and classes, retrieval time is significantly reduced, which is a key advantage during troubleshooting scenarios, where swift and accurate decisions are essential to prevent or mitigate accidents.

In structural and mechanical design, segmentation offers a powerful solution to the common technical contradiction of reducing weight without compromising strength. Heavy components can be segmented into lighter parts connected by joints. A classic example is transforming a solid beam into a lightweight truss composed of straight members connected through joints. Similarly, components can be segmented into functional zones responsible for carrying loads. This technique is employed in the construction of lightweight bridges and foundations. In the construction of foundations, for example, massive monolithic foundations are replaced with multiple lighter foundation segments that collectively provide the required structural support. Segmentation is also employed in product packaging, where instead of using heavy monolithic boxes, only edge segments are used for protection, connected by light links.

6.3 PREVENTING DAMAGE ACCUMULATION AND LIMITING DAMAGE PROPAGATION BY SEGMENTATION

6.3.1 Creating Barriers Containing Damage

Segmentation that prevents damage accumulation and damage propagation enhances reliability and reduces risk through the following mechanisms: (i) *by creating barriers that prevent damage accumulation*, (ii) *by creating barriers preventing the spread of damage*, and (iii) *by introducing deliberate weak boundaries that dissipate or deflect damage*.

Many failures occur when damage accumulates beyond a critical threshold, such as in the case of fatigue failure. By segmenting a system, the accumulation of damage can be delayed, effectively reducing the likelihood of reaching a point where damage begins to propagate. Furthermore, the strong physical boundaries created by segmentation serve to halt the propagation of damage, offering an additional layer of protection. These boundaries can block the pathways through which damage might spread or prevent damage from reaching a critical point. In the event of failure, segmentation can isolate and contain the damage, limiting its impact.

One practical example of the protective effect of segmentation is space segmentation in hazardous environments, such as the use of blast walls to mitigate the impact of blast waves. This is an effective strategy for preventing domino-type failures in situations where fuel tanks or pressure vessels are near each other. The barriers formed by segmentation provide passive protection against the spread of fire, radiation, toxic substances, or other dangerous conditions. Moreover, segmentation can prevent the accumulation of hazardous materials, such as snow, water, or overheated steam, by physically dividing them into smaller, manageable quantities. This reduces the likelihood of catastrophic failures caused by the sudden release of large amounts of stored energy.

Crack arrestors, such as strips or rings made from tougher materials, serve as an example of segmentation that limits the spread of damage by creating strong boundaries. The crack arresting mechanism works by reducing the strain energy flow to the crack tip when it encounters the tougher material, effectively halting crack propagation. A similar concept is applied to minimize the consequences from buckling of pipelines under high external hydrostatic pressure. While increasing the thickness of the pipeline could eliminate buckling, this approach is costly. A more cost-effective and preferable solution is to use buckle arrestors. These are thick steel rings, welded or attached at regular intervals along the pipeline, designed to arrest the propagating buckle and contain damage within a small section. By limiting the extent of buckling, the affected section can be cut out and replaced without the need to overhaul the entire pipeline. The optimal spacing between buckle arrestors is determined through a cost-benefit analysis, balancing the installation cost with the potential repair expenses. In this case, segmentation reduces risk by minimizing the consequences of failure.

Segmentation that limits the spread of damage by creating strong boundaries is widely applicable for reducing technical risk across various domains. One prominent example is in shipbuilding, where the hull is divided into watertight compartments. This type of segmentation ensures that if flooding occurs, only the affected compartments are impacted, allowing the ship to maintain buoyancy. Similarly, space segmentation using fireproof doors in building corridors is a standard practice to delay the spread of fire, thus reducing the risk and impact of fire-related damage.

In the realm of public health, segmenting large groups of people into smaller, isolated units helps prevent the spread of infectious diseases. This concept also applies to the protection of valuable assets, where access is limited through segmentation to reduce the likelihood of theft or damage.

In cybersecurity, segmentation plays a critical role in limiting the spread of damage across computer networks. In a segmented network, gaining access to one segment does not automatically grant an attacker access to others. The segmented design significantly slows down an attacker's ability to reach valuable services and increases the chances of early detection. Additionally, securing each network segment with firewalls adds layers of protection, requiring attackers to breach multiple security barriers before accessing sensitive systems. As a result, segmentation drastically reduces the likelihood of unauthorized access.

6.3.2 Creating Weak Interfaces Dissipating or Deflecting Damage

Segmentation also limits damage through the mechanism of weak interfaces. The weak boundaries created by segmentation often help confine damage once failure has occurred. Thus in a segmented pipeline, a crack is arrested at the edge of the pipeline section because the two segments are simply placed in contact, creating a weak interface. This interface absorbs and dissipates the crack's energy, preventing it from propagating into the adjacent segment. As a result, segmenting the pipeline into multiple parts with weak interfaces confines the damage to a single segment, significantly reducing the consequences of failure.

Similarly, in composite materials, segmentation into multiple layers with weak interfaces can deflect propagating cracks. As cracks encounter these weak boundaries, their progression into the next layer is delayed, thus enhancing the material's fatigue life.

6.3.3 Reducing Deformations and Stresses by Segmentation

Thermal deformations and stresses can be reduced through segmentation, where segments are separated by expansion gaps. These gaps, such as those between rail segments or building panels, accommodate thermal expansion, helping to absorb strains and preventing the formation of large thermal stresses that could otherwise damage the structure.

Similarly, excessive deformation can be minimized by segmenting high-energy processes into smaller, intermediate steps, each associated with smaller energy. This type of segmentation enhances control, reduces intermediate deformations, and minimizes overall distortion, thereby lowering the risk of failure. For instance, instead of applying a large thermal load with a single weld bead, the weld can be divided into multiple smaller, multi-pass welds. Each smaller weld introduces smaller amount of thermal energy, significantly reducing geometric distortion compared to a single weld.

6.3.4 Reducing Hazard Potential by Segmentation

Segmentation of hazardous substances is an effective strategy for limiting the amount of energy stored in these substances, thereby reducing their potential to cause harm. For example, processing small, segmented volumes of toxic substances at a time significantly reduces the risk of poisoning in the event of accidental spillage (Figure 6.2).

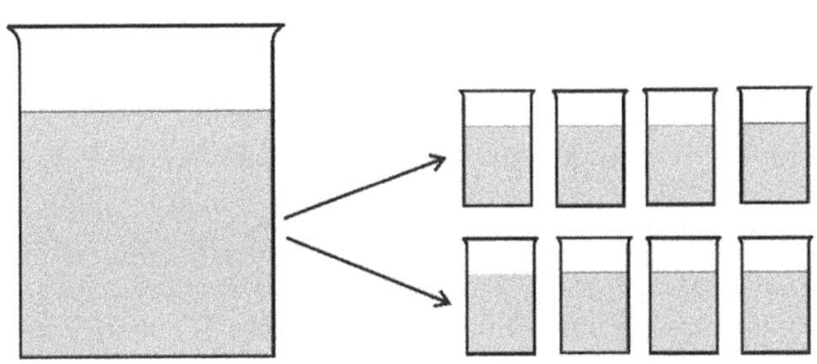

FIGURE 6.2 Segmentation of a toxic substance to minimize the impact of potential spillage.

The emergence of modular nuclear reactors, particularly in response to recent high-profile nuclear disasters, illustrates a shift in design philosophy aimed at improving safety. By reducing the size of reactors from over 700 MW to approximately 300 MW, the potential for catastrophic failure is limited. In the event of an incident, the consequences are contained, thereby reducing environmental impact.

This approach to risk management emphasizes containing damage, rather than attempting to predict the likelihood or the time of a disaster, which is an inherently uncertain and nearly impossible task. This approach aligns with the ‘precautionary principle’ in risk management (Sunstein, 2002), as it avoids steps that could lead to harm. What makes segmentation particularly attractive is that it achieves risk reduction within the existing system, without the need for external interventions. This means that risk is reduced without relying on auxiliary resources, making the method both cost-effective and economically viable.

By limiting the quantity of hazardous material handled at any given time, segmentation eliminates the need for significant investment while ensuring that the potential harm is kept to a minimum. Unlike some risk mitigation strategies, segmentation is a low-cost solution that does not strain safety budgets or lead to prohibitively large expenses.

6.3.5 Reducing the Likelihood of Errors by Segmenting Operations

Segmentation of an operation or design into smaller, incremental steps is an effective strategy for reducing the likelihood of errors. A design approach that breaks down the process into clearly defined steps, each thoroughly assessed and tested before moving to the next, provides better control and makes it easier to identify the source of errors. When an error is identified, the operation or design can be rolled back to the previous stage, where the error was not yet introduced. Adding smaller, incremental steps one at a time allows for pinpointing the exact step at which the error occurred. This approach is the foundation of incremental design, which is especially effective in developing complex software algorithms.

The principle of segmentation is particularly useful in reducing errors when evaluating complex expressions. By segmenting the expression with temporary variables, the complexity is reduced significantly, making it easier to calculate and less prone to mistakes.

6.3.6 Limiting the Presence of Flaws by Segmentation

Segmentation into very small sizes (e.g., spinning extremely thin glass fibres) eliminates the presence of imperfections which reduces the likelihood of failure. This is one reason for the exceptional strength of materials like glass fibres and carbon fibres.

Glass fibres, spun from molten glass, have diameters ranging from 10 to 100 microns. The small cross-sectional size helps eliminate imperfections, contributing to their remarkable tensile strength. For this reason, glass fibres are commonly used as reinforcement in glass fibre polymers.

Similarly, carbon fibres, which have diameters of less than 10 microns, also exhibit outstanding tensile strength due to their small size and the absence of imperfections. These fibres are often woven into textiles and used to reinforce polymers, metals, or carbon matrices (Ashby and Jones, 2002).

6.4 IMPROVING FAULT TOLERANCE AND REDUCING VULNERABILITY TO A SINGLE FAILURE BY SEGMENTATION

6.4.1 Improving Fault Tolerance by Segmentation

A common failure mechanism is present when damage is locally initiated, which then spreads, leading to the complete collapse of a component or structure. Segmentation can be used to enhance resistance to localized damage, preventing it from spreading and causing catastrophic failure.

For example, a monolithic glass panel will shatter completely if struck by an object because the crack created by the impact quickly spreads across the entire panel. However, by segmenting the glass panel into smaller sections, the panel becomes more resistant to localized damage. In this case, only the impacted segment will suffer damage, preventing the crack from propagating into adjacent segments. As a result, the localized failure of a glass segment will not lead to the collapse of the entire panel.

This principle can be further optimized by selectively segmenting only the regions most likely to experience impact. For instance, when a brittle prismatic object falls to the ground, the likelihood of it landing on a corner is significantly higher than the likelihood of it landing flat on one of its sides. As a result, cracks that initiate at the corners are more likely to propagate throughout the object. If only the corners of the prismatic object are segmented, cracks that form at the corners will not spread to the rest of the object.

A similar argument regarding the fault tolerance of segmented structures applies to components subjected to tension (Figure 6.3a, b). Ropes and cables are typical examples of segmented objects, composed of multiple twisted strands or wires. In Figure 6.3a, a monolithic solid rod is loaded with a force with magnitude F. In Figure 6.3b, the rod is segmented into parallel segments, each with a smaller cross-sectional area, such that the total cross-sectional area of the segments equals the cross-sectional area of the monolithic rod.

In the monolithic rod, a flaw can initiate a crack that propagates through the entire cross-section, leading to failure. In contrast, in the segmented design (Figure 6.3b), a fracture in a single segment remains localized and does not propagate into adjacent segments. Thus, even though the material and total cross-sectional area remain unchanged, the segmented rod exhibits significantly greater fault tolerance.

Segmentation transforms a single critical failure at the macro level into multiple non-critical failures at the micro level. In the solid rod, a single critical flaw can lead to sudden catastrophic failure at the macro level. In contrast, in the segmented rod failure occurs gradually, through non-critical failures of multiple segments.

A closely related example involves the failure behaviour of a solid steel rod compared to that of a segmented cable of equivalent strength, constructed by twisting many thin wires. Instead of abrupt collapse, the cable undergoes progressive

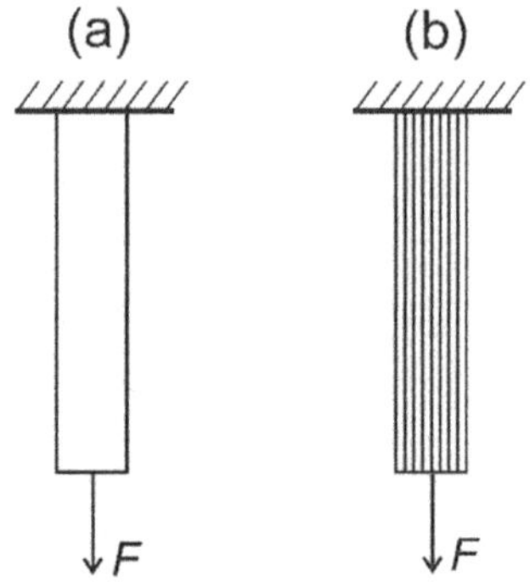

FIGURE 6.3 (a) Monolithic rod; (b) segmented rod.

deterioration as individual wires fail over time. Routine inspection can detect when the number of failed surface wires exceeds a predefined threshold. At that point, the cable can be proactively replaced, thereby avoiding catastrophic consequences associated with sudden failure.

6.4.2 Case Study: Reducing the Vulnerability to a Single Failure of a Cooling System by Segmentation

Consider a scenario in which a single pump is used to cool a chemical reactor (Figure 6.4a). If the pump fails, the cooling process is interrupted, potentially leading to catastrophic consequences, production delays, and significant losses.

Now suppose the single pump is replaced by four smaller pumps, collectively providing the same total flow (Figure 6.4b). In this segmented configuration, failure of one or even two pumps does not lead to a catastrophic shutdown; the system continues to operate maintaining a safe level of cooling for the reactor. In this case, segmentation reduces risk by replacing a single-point failure with a series of non-critical failures. It essentially transforms a sudden, system-wide failure into gradual degradation, allowing time for detection, intervention, and repair.

An analytical justification of this principle can be given with the single pump in Figure 6.4a, which has a reliability $r = 0.75$ over one year of continuous operation. Suppose the pump is segmented into four smaller pumps, each with a smaller individual reliability $s = 0.70$ over the same period. The system is designed in such a way that any two of the four pumps are sufficient to maintain a safe level of cooling. The probability that the segmented cooling system will operate continuously for one year without requiring a shutdown due to insufficient cooling is equal to the probability that at least two of the four pumps (i.e., two, three, or all four) remain functional over that period.

Assuming the pumps fail independently, this probability is given by:

$$R = s^4 + \frac{4!}{3! \times (4-3)!} s^3 (1-s)^1 + \frac{4!}{2!(4-2)!} s^2 (1-s)^2 = 0.916 \qquad (6.1)$$

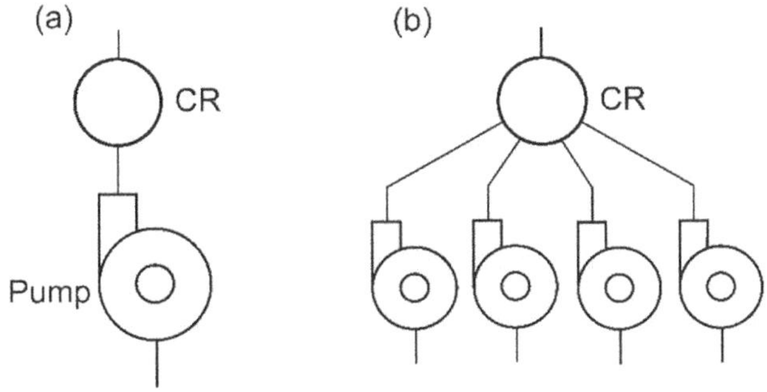

FIGURE 6.4 (a) A single pump cooling a chemical reactor (CR); (b) multiple pumps cooling the chemical reactor reduces the vulnerability to a single failure and improves fault tolerance.

As a result of segmentation, the reliability of the cooling system increases significantly, from 0.75 to 0.916, even though the individual pumps have lower reliability than the original single pump. This improvement arises from the system's ability to tolerate multiple non-critical failures without functional loss.

A similar principle applies to mechanical assemblies such as flanges. In a flange secured by only a few fasteners, the failure of any single fastener can result in loss of containment because each fastener represents a critical point of failure. Now consider a flange secured by multiple smaller fasteners, each with smaller individual reliability. In this segmented configuration, the assembly is no longer vulnerable to the failure of a single fastener or even several fasteners. The original, macro-level critical failures have been replaced by micro-level non-critical failures, transforming sudden critical failures into gradual deterioration.

The segmentation approach also enhances the reliability of sensing systems that trigger alarms or actions in response to dangerous conditions, such as elevated concentrations of chemicals, excessive force, torque, pressure, temperature, or humidity. Replacing a single high-reliability sensor with multiple lower-reliability sensors improves fault tolerance. The system becomes resilient to individual sensor malfunctions or even to a simultaneous failure of several sensors.

Segmentation reducing the risk of losing an asset is practiced whenever the valuable assets are divided (segmented) into smaller parts, each of which is subsequently stored in a different place. Accidental access to any part of the asset will not result in an automatic loss of the entire asset. Segmenting a single command centre into a mesh-type network of interconnected parts, significantly reduces the vulnerability to strikes.

6.5 REDUCING LOADING STRESSES BY SEGMENTATION

6.5.1 Improving the Load Distribution by Segmentation

Segmenting a component into several smaller parts increases the contact area, which often leads to a significant reduction in loading stresses. This segmentation mechanism enhances reliability by distributing the load across multiple load-bearing units.

Consider a flange with very few fasteners. When the number of fasteners is too small, the load is not evenly distributed, leading to excessive stress on some of the fasteners. By increasing the number of fasteners the load is shared more uniformly which enhances the overall reliability of the assembly.

Similarly, segmenting a component can increase its contact area and improve conformity with adjacent components. This results in significantly reduced contact stresses, which in turn reduces wear and increases resistance to fracture under overload. For instance, using multiple rows of ball bearings instead of a single row distributes the load more effectively, thereby decreasing contact stress and wear on each bearing.

Often, the improved load distribution achieved by segmentation is used to achieve a greater level of balancing. The increased level of balancing results in a greater stability, smaller vibration amplitudes and inertia forces. Such is for example the case of introducing multiple cylinders in an internal combustion engine, multiple blades in a turbine, etc.

6.5.2 Case Study: Segmentation of Concentrated Loads to Improve Reliability and Develop Light-Weight Designs

There are numerous cases where design-engineers have control over the points of application of external loads. For the simply supported beam with length a in Figure 6.5a, the concentrated load F is applied in the middle and results in a bending moment $M(x)$ (Figure 6.5b). The maximum bending moment $M_{1,\max}$ is attained at $x = a/2$ and is equal to $M_{1,\max} = Fa/4$ (Figure 6.5b). Segmenting the concentrated load F into two loads with magnitude $F/2$ (Figure 6.5c) reduces the maximum bending moment three times, from $M_{1,\max} = Fa/4$ to $M_{2,\max} = Fa/12$ (Figure 6.5d). The reduction of the bending moment reduces the bending stress in the beam and increases its resistance to overstress failure.

In some design applications (e.g., in motorsport design) the focus is often on obtaining a light-weight design, not on increasing the resistance to overstress failure. A light-weight design translates directly into enhanced performance, reduced fuel consumption, and reduced emissions. As a result of the segmented external load and the reduced tensile stresses from bending in Figure 6.5c, the cross-section of the loaded beam can be reduced which results in a light-weight design.

Indeed, the maximum bending stress σ_b in a beam with a circular cross-section with diameter d is given by the well-known formula (Gere and Timoshenko, 1999): $\sigma_b = 32M/(\pi d^3)$ where M is the maximum bending moment. Reducing the bending moment three times at the same maximum bending stress σ_b, results in a significant reduction of the cross-section diameter of the beam. From $\sigma_b = 32M/(\pi d^3) = 32(M/3)/(\pi d_1^3)$, the diameter of the light-weight beam is evaluated to be $d_1 = 0.693d$, which, for a uniform cross section, results in volume of the material per unit length of the beam equal to $\pi(0.693d)^2/4 = 0.48 \times \pi d^2/4$. As a result, the light-weight design carries the same maximum bending stress σ_b, with

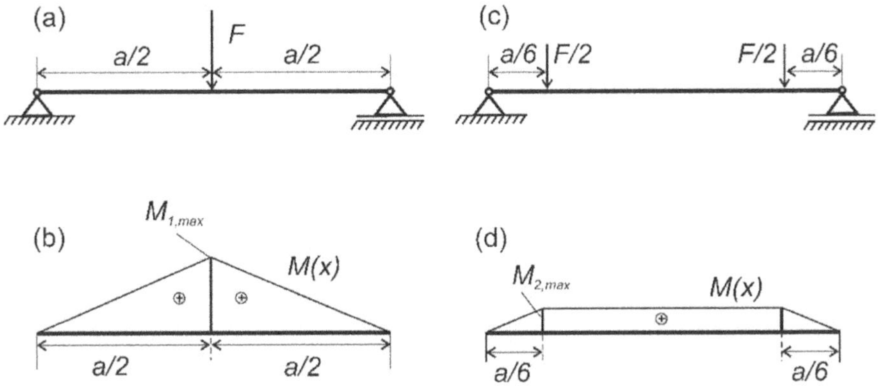

FIGURE 6.5 Reducing the risk of overstress failure of a beam; (a) beam loaded with a single external concentrated load *F*; (b) bending moment diagram associated with a single concentrated load; (c) beam loaded with a segmented external load; (d) bending moment diagram associated with the segmented load.

only 48% of the material of the original beam. The weight saving from segmenting the external loading force is impressive.

Interestingly, this simple solution for reducing the stresses in loaded structures, based on segmentation of external concentrated loads, has never been suggested in standard textbooks in the mature fields of stress analysis and strength of components.

6.5.3 Improving Heat Dissipation by Segmentation

Increasing the surface-to-volume ratio through segmentation enhances the total surface area of the component parts, which facilitates more efficient heat dissipation. Effective heat dissipation is critical for the reliability of many devices, particularly electronic ones. Rapid heat removal and maintaining the equilibrium temperature within acceptable limits are essential for fault-free operation of electronic components.

For a fixed component volume, segmentation increases the available surface area for heat transfer, thereby improving cooling. Enhanced heat transfer results in a lower equilibrium temperature, which in turn reduces the rate of degradation processes and promotes more reliable performance. When direct segmentation of heated components is not feasible, highly segmented radiators can be attached to achieve similar thermal benefits.

Heat dissipation also plays a vital role in the reliability of mechanical systems. For example, replacing a single *V*-belt with multiple parallel *V*-belts improves thermal performance. The segmentation allows for more effective dissipation of heat generated from hysteresis losses, leading to a lower equilibrium temperature in the belt material and improved system reliability.

6.6 REDUCING THE PROBABILITY OF A LOSS/ERROR BY SEGMENTATION

The risk of failure can be reduced by dividing an investment into smaller portions allocated across multiple unrelated sectors with non-correlated returns. While segmentation does not alter the expected return, it significantly reduces the variance (volatility) of the overall portfolio (Teall and Hasan, 2002). Investment funds mitigate risk by purchasing shares in a wide range of companies. Even small investors can achieve a segmented and well-diversified portfolio at low transaction costs by investing in financial vehicles such as unit trusts.

6.6.1 Case Study: Reducing the Likelihood of a Loss by Segmenting Opportunity Bets

Segmentation can be effectively used to reduce the cost of mitigating risk in risk-reward scenarios. Such events can result in either gains or losses. A common example is an investment in a particular enterprise: a successful investment yields returns (benefits), while an unsuccessful one leads to losses.

Suppose that $0 \le p_s \le 1$ is the probability that the risk-reward bet will be a 'success' and $p_f = 1 - p_s$ is the probability that the risk-reward bet will be a 'loss'.

The expected values of the benefit and the loss given that the risk-reward event has materialized are denoted by $\bar{B}_s$ and $\bar{C}_f$, respectively.

The expected profit $\bar{G}$ from a risk-reward bet is then given by:

$$\bar{G} = p_s \times \bar{B}_s + p_f \times \bar{C}_f \tag{6.2}$$

where p_s is the probability of a beneficial outcome with expected magnitude $\bar{B}_s$ and $p_f = 1 - p_s$ is the probability of a loss with expected magnitude $\bar{C}_f$ (the loss $\bar{C}_f$ is a negative value). If $\bar{G} > 0$, the risk-reward bet will be referred to as *opportunity bet*.

The potential profit G from a risk-reward bet is a random variable following a Bernoulli distribution with parameter p_s. For constant values of the benefit given success $\bar{B}_s$ and the loss given failure $\bar{C}_f$, the probability distribution of the potential profit G is given by $P(G = \bar{B}_s) = p_s$ and $P(G = \bar{C}_f) = p_f$.

Consider a risky prospect containing a single opportunity bet with parameters: $p_s = 0.3$, $\bar{B}_s = 300$, $p_f = 0.7$, $\bar{C}_f = -90$ for which the expected profit is $E(G) = \bar{G} = 0.3 \times 300 - 0.7 \times 90 = 27$. The probability of a net loss from this risky prospect is 70%.

The risk of a net loss from this risky prospect can be reduced by segmenting the opportunity bet. This creates a second risky prospect which now contains three opportunity bets with the same probability of success and failure but with three times smaller magnitudes for the benefit given success and the loss given failure: $p_s = 0.3$, $\bar{B}_s = 300/3 = 100$, $p_f = 0.7$, $\bar{C}_f = -90/3 = -30$. The expected profit *from the risky prospect containing the three segmented opportunity bets* is

$E(G_{123}) = \bar{G}_{123} = 3 \times (0.3 \times 100 - 0.7 \times 30) = 27$. Because a net loss from the second risky prospect can be generated only if a loss is generated from every single segmented bet, the probability of a net loss from the segmented bets is $p_{f,123} = 0.7^3 \approx 0.34$.

Clearly, the second risky prospect, featuring a segmented opportunity bet, is preferable to the first. Segmenting the opportunity bet in the initial prospect significantly reduces the associated risk.

This case study demonstrates that the risk of an opportunity bet can be significantly reduced by dividing it into multiple smaller bets, each with the same probability of success and failure as the original, but with proportionally reduced gains and losses. This example also shows that the number of the risk-reward activities should be a key consideration in selecting a risky prospect. More detailed treatment about reducing the risk from opportunity bets can be found in Todinov (2013).

6.6.2 Reducing the Likelihood of Erroneous Conclusion from Imperfect Tests through Segmentation

Another important application of the segmentation method lies in its ability to cost-effectively reduce the likelihood of drawing erroneous conclusions from an imperfect test.

Consider a test on a substance with mass M, which yields a binary outcome: positive or negative, regarding the presence of a specific attribute in the substance. Although the test is imperfect and carries an error probability of p, this probability can be reduced by segmenting the test substance with mass M into n parts and conducting separate tests on the individual parts with mass M/n, rather than relying on a single test of the entire substance with mass M (see Figure 6.6).

Suppose that each test conducted on a segmented portion of the substance has a probability p of producing an error. Once the results from all tests are available, the presence of the attribute is determined by the majority outcome (positive or negative) across the individual tests.

For the segmented test to yield an incorrect conclusion, more than half of the tests must produce erroneous results. In the special case of $n=2k+1$ identical, independent tests, this means that at least $k+1$ of them must be incorrect. Since the tests are statistically independent and the probability of an erroneous test is constant, the process satisfies the conditions of a binomial experiment (DeGroot, 1989). Therefore, the number of erroneous outcomes, denoted by X, follows a binomial distribution.

The probability that the number of errors is greater than or equal to $k+1$ is given by:

$$P(X \geq k+1) = \sum_{x=k+1}^{n} \frac{n!}{x!(n-x)!} p^x (1-p)^{n-x} \tag{6.3}$$

For $n=11$ and $p=0.1$, for example, the probability of erroneous output from the segmented tests is

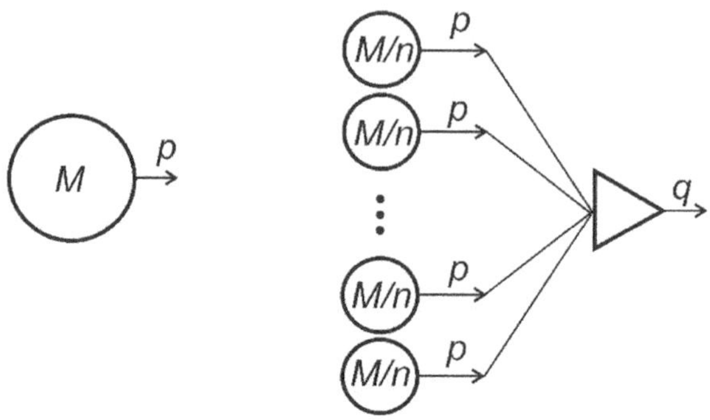

FIGURE 6.6 Segmenting a substance of mass M into n smaller parts (M/n) and conducting multiple parallel tests on the segments, drastically reduces the probability of erroneous conclusion from the tests.

$$P(X \geq 6) = \sum_{x=6}^{11} \frac{11!}{x!(11-x)!} 0.1^x (1-0.1)^{11-x} = 0.0003 \tag{6.4}$$

As a result, the relatively high probability of error $p=0.1$ associated with a single test has been reduced by a factor of 333 through segmentation.

The same segmentation principle can be applied to improve decision-making processes, policies, transactions, or other critical actions. By distributing the decision across multiple independent decision-makers, the likelihood of reaching an incorrect conclusion can be significantly reduced.

For a domain expert, such as a specialist in food safety or toxicology, it may not be immediately apparent that segmenting the test sample can so drastically reduce the risk of an error. Importantly, this method offers an effective approach to risk reduction, as it avoids the need for more advanced or precise testing equipment that reduces the error in a single test.

6.7 IMPROVING RELIABILITY AND REDUCING RISK BY A SEGMENTATION THROUGH SUB-ADDITIVE AND SUPER-ADDITIVE ALGEBRAIC INEQUALITIES

There exist algebraic inequalities whose meaningful physical interpretation yields superior performance of a system or process by a segmentation of a controlling factor. Inequalities based on sub-additive and super-additive functions can be used for this purpose.

Consider a factor x that has been segmented into a number of non-negative parts $x_1, x_2, ..., x_n$ ($x = x_1 + x_2 + ... + x_n$); $n \geq 2$. The output associated with the factor x and each of the parts x_i is given by the function $f(x)$. The output function $f(x)$ is said to be *sub-additive* if for any set $x_1, x_2, ..., x_n$, the next inequality holds:

$$f(x_1 + x_2 + ... + x_n) \leq f(x_1) + f(x_2) + ... + f(x_n) \quad (6.5)$$

The output function $f(x)$ is said to be *super-additive* if, for any set $x_1, x_2, ..., x_n$, the next inequality holds:

$$f(x_1 + x_2 + ... + x_n) \geq f(x_1) + f(x_2) + ... + f(x_n) \quad (6.6)$$

Note that if the output function $f(x)$ is linear, equality is attained in inequalities (6.5) and (6.6). Multiplying the definition inequalities (6.5) and (6.6) by '-1', reverses the direction of the inequalities. Consequently, if $f(x)$ is sub-additive, $-f(x)$ is super-additive and if $f(x)$ is super-additive, $-f(x)$ is sub-additive.

A central property related to sub-additive and super-additive inequalities can be stated. If a function $f(x)$, with a domain $[0, \infty)$ and range $[0, \infty)$ is concave, then the function is sub-additive (see inequality 6.5). If the function $f(x)$ is convex, and if $f(0) \leq 0$ then the function is super-additive (see inequality 6.6). The proof of these properties can be found in Todinov (2025a).

The sub-additive inequality (6.5) and the super-additive inequality (6.6) have very important potential applications. If the function $f(x)$ measures the effect/output of a particular factor and x_i denote the different magnitudes of an additive factor, inequalities (6.5) and (6.6) *provide the unique opportunity to increase the effect of the factor by segmenting or aggregating it, depending on whether the function $f(x)$ is concave or convex*. If the function is concave, with a domain $[0, \infty)$ and range $[0, \infty)$, segmenting the factor results in a larger output. If the function is super-additive, with a domain $[0, \infty)$ and range $[0, \infty)$, aggregating the factor results in a larger output.

Knowing whether the function is sub-additive or super-additive is sufficient to reduce uncertainty and make a correct decision, irrespective of the magnitudes x_i of the factors characterizing the different segments.

Let the function $f(x)$ describe the output of a physical process and x stand for a particular additive factor on which the output quantity depends. Let the concave function $f(x)$ be given by the concave power law

$$f(x) = kx^p \quad (6.7)$$

where $p < 1$ is the exponent of the power law and $k > 0$ is constant.

The concave power law (6.7) is widely used in science and engineering to describe the output of physical processes that experience saturation. Below are several examples of processes where concave power laws provide a good approximation:

- The increase in a material's strength or other properties as a function of the added reinforcing substance often follows a concave power law due to saturation effects.
- Because of limited solubility, the growth of gas solubility in a liquid as a function of the gas's partial pressure can often be well approximated by a concave power law.

- The rate of extraction of resources like oil or minerals frequently follows a concave power law, as remaining reserves become progressively harder and more costly to access.
- In mature economies, growth often exhibits concave power law behaviour due to diminishing returns on investment and resource limitations.
- The development of skill or knowledge over time or with practice typically follows a concave power law, as individuals near their performance ceilings.

Even when a function is not a strict concave power law, many concave functions can still be closely approximated by one.

6.7.1 Case Study: Segmentation of a Filter to Reduce Pollution

According to Fu et al. (2010), the dependence of the collection efficiency and thickness of the fibrous media in the filter, for a specified duration, is described by a nonlinear, concave curve reaching a region of saturation with increasing the thickness of the filter. A plot collection efficiency versus log-thickness based on experimental results published in Fu et al. (2010) can be fitted very well with a straight line (Figure 6.7) which shows that in the range 8–300 μm, the trend 'Collection efficiency – Thickness of filter in μm' can be approximated very well by the concave power law (Figure 6.8) with power approximately equal to $p \approx 0.27$.

The quantity of the absorbed substance versus thickness of filter therefore is given by a concave power law (Figure 6.8). The concave dependence in Figure 6.8 can be exploited to increase the performance of such filters through reverse engineering of an algebraic inequality.

Let the quantity $f(x)$ describe the collection efficiency of the filter and x stand for the thickness of the fibrous media. Let the concave function $f(x)$ be approximated, by the concave power law (6.7).

Consider the algebraic inequality

$$k(x_1 + x_2 + \ldots + x_n)^p < kx_1^p + kx_2^p + \ldots + kx_n^p \tag{6.8}$$

which always holds if $p < 1$ and $x_i > 0$. The proof of inequality (6.8) can be found in Todinov (2024a).

According to inequality (6.8), the quantity of absorbed harmful substances can be increased by segmenting the filter. This is because of the concave dependence (6.8) (Fu et al., 2010).

Segmenting the filter with thickness s of the fibrous absorbing media, into smaller sections with thicknesses $s_1, s_2, \ldots, s_n$ whose sum is equal to the same total thickness s of the fibrous media (Figure 6.9) ($s_1 + s_2 + \ldots + s_n = s$), yields a greater collection efficiency and smaller pollution. The quantities f_i of absorbed substance by the segments are given by

$$f_1 = ks_1^p,\ f_2 = ks_2^p, \ldots, f_n = ks_n^p \tag{6.9}$$

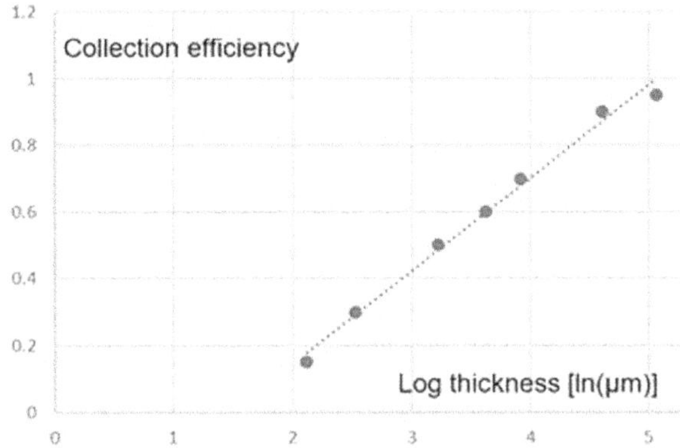

FIGURE 6.7 Collection efficiency versus log-thickness of a filter according to experimental measurements (Ciupitu, 2018).

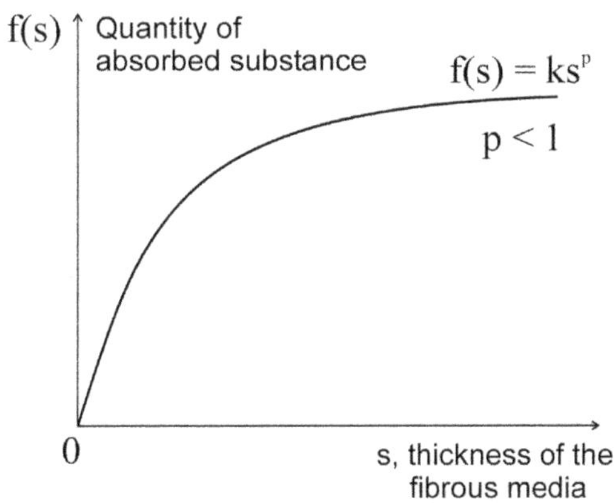

FIGURE 6.8 Quantity of absorbed substance (collection efficiency) versus thickness of the fibrous media in the filter, approximated by a concave power law.

and their sum results in a larger total quantity of absorbed substances because of the algebraic inequality:

$$k(s_1 + s_2 + \ldots + s_n)^p < ks_1^p + ks_2^p + \ldots + ks_n^p \tag{6.10}$$

Note that if the collection efficiency of the filter were linear, there would not be any benefits for the collection efficiency. Indeed for $p = 1$, inequality (6.10) transforms into equality:

$$k(s_1 + s_2 + \ldots + s_n) = ks_1 + ks_2 + \ldots + ks_n \tag{6.11}$$

Note that if the collection efficiency of the filter were convex (bending upwards), there would also be no benefits for the collection efficiency. In fact, there would be a reduction in the collection efficiency due to the segmentation. Indeed, in this case, $p > 1$, and

$$k(s_1 + s_2 + ... + s_n)^p > ks_1^p + ks_2^p + ... + ks_n^p \tag{6.12}$$

6.7.2 Case Study: Reducing the Risk of Formation of Undesirable Brittle Phase by Segmentation

Consider another case study related to formation of undesirable brittle phase during cooling of a spherical workpiece with volume V, starting from a specific initial temperature. The amount of the brittle phase, produced during cooling, follows a power-law relationship, expressed as:

$$z = aV^p \tag{6.13}$$

where a and p ($p > 1$) are constants which depend on the initial temperature, the nature of the material, and the shape of the workpiece. The controlling factor in the power-law dependence (6.13) is volume V of the workpiece.

The formation of a brittle phase during cooling degrades the mechanical properties of the specimen, as increased amount of brittle phase leads to a significant reduction in strength. If, during processing, there is an option to use either a single workpiece of volume V or multiple workpieces whose total combined volume is also V, the key question is: which configuration results in a smaller total amount of brittle phase formed during cooling?

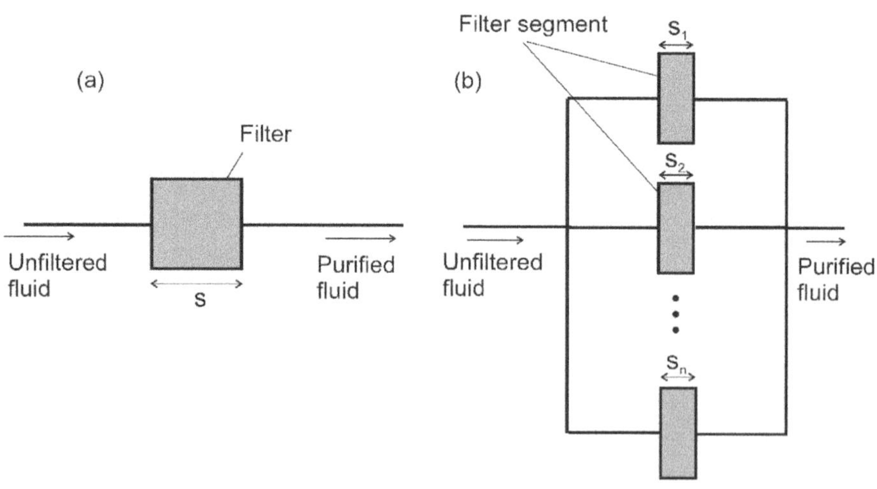

FIGURE 6.9 (a) Original and (b) segmented filter with greater collection efficiency.

Since $p > 1$, the quantity z of brittle phase is a convex function of the volume of the workpiece because the second derivative with respect to the volume V is positive: $d^2z/dV^2 = ap(p-1)V^{p-2}$. Because for a volume varying in the interval $[0, \infty)$ the quantity of brittle phase varies within the range $[0, \infty)$ and $z(0) = 0$, the function (6.13) giving the quantity of brittle phase is super-additive. Consequently, the inequality

$$a(V_1 + V_2 + ... + V_n)^p > aV_1^p + aV_2^p + ... + aV_n^p \tag{6.14}$$

holds for $V_i > 0$. The right side of inequality (6.14) can be physically interpreted as the sum of the quantities of brittle phase formed during cooling of n workpieces with volumes $V_1, V_2, ..., V_n$. The left side of inequality (6.14) can be physically interpreted as the quantity of brittle phase formed during cooling of a single workpiece with a similar shape and volume equal to the sum of the volumes $V_1, V_2, ..., V_n$ of the smaller workpieces.

As a result, from inequality (6.14), it follows that a segmentation of a workpiece with volume V into n workpieces with similar shape and volumes $V_1, V_2, ..., V_n$, $(V = V_1 + V_2 + ... + V_n)$ whose sum is equal to the volume of the single workpiece, decreases the total amount of unwanted brittle phase during cooling.

7 Reducing the Risk of Failure by Deliberate Weaknesses

7.1 REDUCING RISK BASED ON INTRODUCING DELIBERATE WEAKNESSES

This chapter focuses on an important domain-independent method for risk reduction based on introducing deliberate weaknesses or repositioning existing weaknesses. The deliberate weaknesses *are deliberately created weak spots towards which a potential failure is channelled. By channelling failure into weak spots, designed to fail in a predictable way, the consequences from failure are reduced.* Should unfavourable conditions occur, the deliberate weakness is the one to fail and protect the expensive parts of the system. In this way, the losses from failure are limited. In the risk equation introduced in Chapter 1:

$$K = p_f C \tag{7.1}$$

where p_f is the probability of failure and C is the cost given failure, the deliberate weaknesses reduce the risk K by reducing the cost C given failure.

Another significant advantage is that the deliberate weaknesses are maintenance-free and constantly ready to operate.

Risk reduction by introducing weak links and stress limiters has already been used extensively in engineering for preventing the stresses from reaching dangerous levels (Eder and Hosnedl, 2008). Familiar examples of deliberate weaknesses are the electrical fuses and circuit breakers, the crumple zones in road cars, the crash cones in racing cars, the shear pins, the sacrificial anodes, and the rupture disks in pressure vessels.

Although the deliberate weak links have been used widely in engineering, their discussion in the reliability and risk literature is very limited, restricted around the few very well-known applications mentioned earlier. Thus, in Pahl et al. (2007), deliberate weak links have only been mentioned as protective devices in discussing indirect safety. The discussion of deliberate weak links in Booker at al. (2001) has been limited to a discussion of the design of shear pins in transmission shafts only.

DOI: 10.1201/9781003679301-7

Despite the existence of numerous applications of deliberate weaknesses in design, the different categories and classes of deliberate weaknesses and the mechanisms through which the deliberate weaknesses from each category reduce risk have only been recently discussed in Todinov (2020). The engineering design literature is concerned mainly with reducing risk by removing weaknesses, not by creating or repositioning weaknesses. A common reliability allocation strategy in engineering design is to make all parts with comparable reliability or strength and not to leave deliberate weak links. Such is, for example, the AGREE (Advisory Group on Reliability of Electronic Equipment) methods described in Ebeling (1997) and the maximum 'distance' to the constraint strategy described in Thomson (1999).

Nonetheless, the importance of reducing risk by implementing deliberate weaknesses is increasing. The modern tendency towards lightweight designs requires reducing to a minimum the cross-sections and the factors of safety of components. As a result, components are fully loaded which results in working stresses that are close to the critical stresses triggering failure. Compared to the old and heavy designs with large safety factors, the modern lightweight designs have small safety factors and the critical stresses can be exceeded even by a moderate overload. Furthermore, while old design codes are based on static loading, modern design codes are also capturing the effects of dynamic loading and the effects from sudden application of loads which are associated with a great deal of uncertainty. For example, it is a well-documented fact that the stresses from sudden application of the load are often twice the stresses from slow, static load application (Gere and Timoshenko, 1999).

Introducing deliberate weaknesses is an important barrier assuring the protection of valuable entities from damage due to excessive stresses. Consequently, this chapter introduces a domain-independent method for reducing the consequences of failure based on deliberate weaknesses, proposes a detailed classification of the categories of deliberate weaknesses that can be used for risk reduction and reveals the mechanisms through which the deliberate weaknesses reduce risk. The chapter also demonstrates that combining the domain-independent method of deliberate weaknesses with domain-specific knowledge results in effective risk reduction. Finally, the drawbacks of the equal-reliability concept are discussed.

7.2 CLASSIFICATION OF TECHNIQUES FOR REDUCING RISK BY CREATING DELIBERATE WEAKNESSES

The techniques for reducing risk by creating deliberate weaknesses can be classified broadly into the following categories (Figure 7.1):

- *Repositioning the existing weaknesses;*
- *Protecting from excessive stresses;*
- *Preventing excessive damage;*
- *Deflecting damage from places where the cost of failure is high;*
- *Providing warning;*
- *Providing rapid access;*
- *Providing quick escape;*
- *Triggering protection systems.*

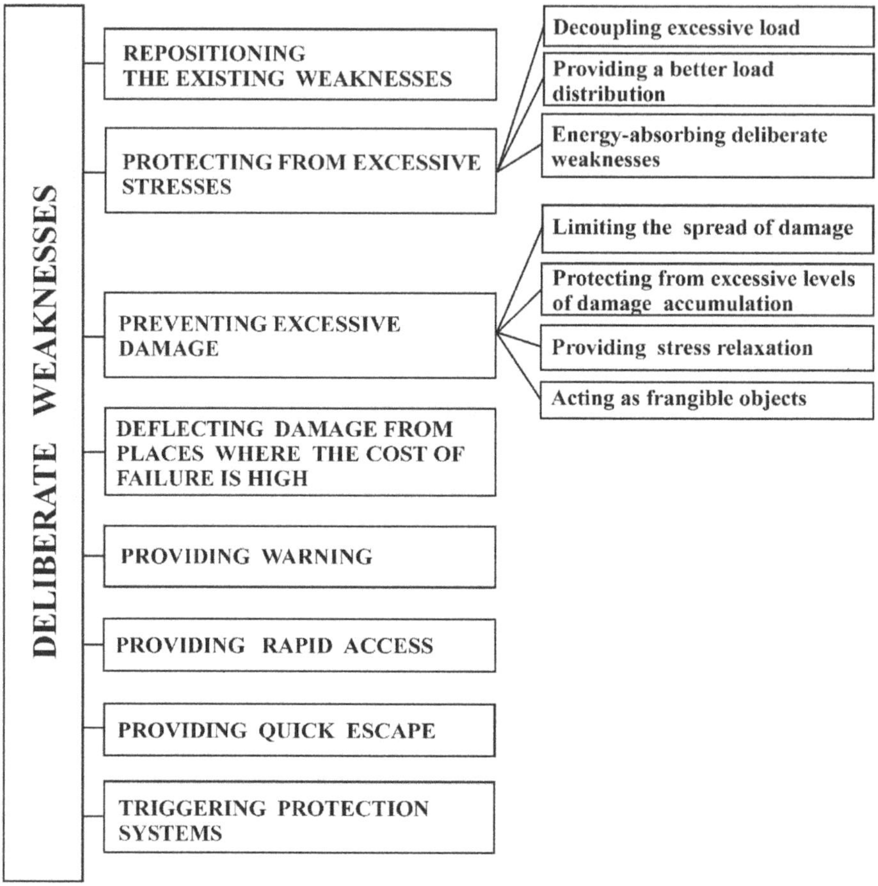

FIGURE 7.1 Classification of deliberate weaknesses for reducing risk.

The techniques outlined in the classification have been distilled from a large number of engineering solutions, each of which was analysed to assess recurring risk reduction patterns. In what follows, a description of the mechanisms through which the techniques from each category reduce risk is provided.

7.3 DELIBERATE REPOSITIONING OF AN EXISTING WEAKNESS TO REDUCE THE CONSEQUENCES OF FAILURE

The essence of this category consists of a deliberate re-positioning of an existing weakness to minimize the consequences of failure. To some extent, this category is related to a category described later which consists of introducing a deliberate weakness which deflects the failure location to a place where the cost of intervention is minimal. The difference is that for the category described in this section, no deliberate weakness is created but an existing weakness is repositioned in time or space with the purpose of reducing the consequences from failure.

Suppose that a process consists of a number of independent operations that can be executed in any order. To obtain a result from the process, all operations must be executed. If any of the operations fail to complete successfully, the execution of all operations in the process is initiated again.

Through a deliberate repositioning in time of the weakest operation, characterized by the highest probability of failure, an early failure of the weakest operation will make it unnecessary to execute the rest of the operations and will reduce significantly the waste of resources on repeating already executed operations. Consequently, the weakness (the operation characterized by the highest probability of failure) must be repositioned by placing it first in the order of executing the set of operations. The effectiveness of this approach has been proved in Todinov (2020) by using both analytical reasoning and Monte Carlo simulations.

7.4 DELIBERATE WEAKNESSES PROTECTING FROM EXCESSIVE STRESSES

A deliberate weakness can be created easily by rating the deliberate weakness at the level of stress at which its failure must be triggered and by eliminating safety margins. Leaving safety margins will make the behaviour of the deliberate weakness unpredictable. Due to the safety margin, the deliberate weakness may not fail and protect the valuable part of the system. Almost any physical property can potentially be used for creating a deliberate weakness: shear strength, yield strength, electrical resistance, electrochemical potential, magnetic properties, melting point, stiffness, fracture toughness, etc.

The shear pin is an example where the shear strength of the material is used to create a deliberate weakness. The thermal fuse is an example where the melting point of an alloy is used for this purpose and the sacrificial anode is an example where the electrochemical potential of a metal/alloy is used.

Weak links in towing cables are safety devices designed to protect towing equipment, such as winches, tow bars, aircraft nose gear, or vehicle drivetrains, from catastrophic overload if towing forces suddenly spike. A weak link is a short, replaceable section of the towing line or connector engineered with a controlled breaking strength slightly below the safe load limit of the rest of the system.

In aviation, glider towlines often incorporate weak links rated to fail before the towplane reaches its structural load limit.

In maritime and industrial winching, weak links may be:

- Softer metal shackles with a reduced cross-section
- Synthetic rope sections with calibrated tensile ratings
- Pre-notched or grooved shear pins

When a weak link fails, the towing operation stops instantly, indicating a dangerous overload. Many weak links are brightly coloured or tagged to allow quick post-failure identification.

By sacrificing the weakest component, the system avoids costly damage such as:

- Bent tow bars
- Sheared bolts in nose gear assemblies
- Broken winch drum shafts

It is important to emphasize that the design of a deliberate weakness requires examining its failure modes and their causes. A fault tree is a suitable structure for this purpose. The basic function of a deliberate weakness is to fail at the specified level of stress. Negating this function gives two fundamental types of faults of deliberate weaknesses: (i) the deliberate weakness fails at stress significantly below the specified level and (ii) the deliberate weakness fails at stress significantly above the specified level.

Consider failure of a shear pin significantly above the specified level of stress. Possible contributing reasons for this type of failure are: (i) large variation of the shear strength of the material; (ii) inhomogeneous material, causing a local fluctuation of the shear strength in the shear zone; (iii) bending of the pin instead of shearing; (iv) incorrect shape and dimensions of the shear pin; (v) incorrect material of the shear pin; (vi) ageing of the material of the shear pin associated with second-phase precipitations increasing strength.

7.4.1 Deliberate Weaknesses Decoupling Excessive Loads

The mechanism behind this category of deliberate weaknesses is preventing excessive stress causing damage by decoupling the load from the structure/component.

Thermal fuses, for example, decouple excessive levels of temperature. Electrical fuses and circuit breakers are familiar examples of deliberate weaknesses decoupling excessive currents while surge protection devices decouple excessive voltage. Shear pins and torque wrenches decouple excessive forces and torques while pressure relief valves and rupture disks decouple excessive pressure.

Many stress-limiting devices compete with the decoupling deliberate weaknesses in mitigating the consequences from failure. It needs to be pointed out that a large part of the stress-limiting devices is, in fact, active protection systems. They may include a sensor, a control unit and a device which disconnects the protected part of the system or limits the stress to which it is subjected. This introduces complexity and potential failure modes which decrease the reliability of active protection systems. In some critical cases (associated with high consequences of failure), a deliberate weakness is still needed as a back-up of the active protection system. Such is the case with the active protection system monitoring the pressure increase in a pressure vessel and activating a pressure-release valve if the pressure exceeds a certain critical limit. Despite the active protection system in place, a deliberate weakness such as a rupture disk is still needed as a back-up. Rupture disks have become a necessary part of the design codes of pressure vessels.

Unlike the active protection systems, solutions based on deliberate weaknesses are very simple, maintenance-free, and characterized by a very high reliability. As a

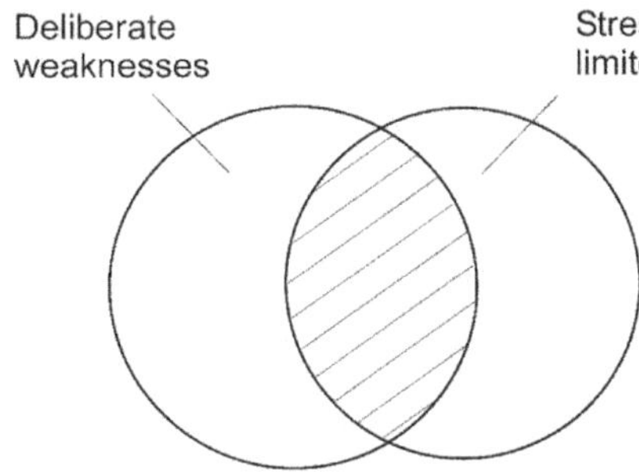

FIGURE 7.2 Deliberate weaknesses and stress limiters.

result, the deliberate weaknesses provide additional, more secure layer of protection against the consequences from failure.

It needs to be pointed out that there is a difference between deliberate weaknesses and stress limiters. While many deliberate weaknesses act as stress limiters, not all deliberate weaknesses are stress limiters, and not all stress limiters are deliberate weaknesses (Figure 7.2).

Deliberate weaknesses providing only warning about stress exceeding a dangerous limit, for example, are not necessarily stress limiters. Conversely, the screw with a shoulder (Erhard, 2006) designed to prevent over-tightening of plastic materials is a stress limiter without being a deliberate weakness.

7.4.2 Deliberate Weaknesses Providing Better Load Distribution

The mechanism behind this category of deliberate weaknesses is the deliberate addition of an extra degree of freedom to ensure that a better balance of forces is obtained. Adding extra degrees of freedom often helps redistribute the load so that each support receives an equal portion of the load and overloading of supports is avoided. Introducing an extra degree of freedom also helps the alignments of parts and reduces stresses. Common examples are the self-aligning bearings (Figure 7.3) where the reduction of the contact stresses at the edges (Figure 7.3a) is achieved by introducing an interface providing extra degrees of freedom (Figure 7.3b).

Statically indeterminate structures also benefit from this type of deliberate weaknesses which help reduce the internal stresses. Introducing an elastic mounting of parts as opposed to a rigid mounting is also an example of a deliberate weakness which helps the alignment of parts, the load redistribution, and stress reduction.

Adding a hinge reduces bending stresses by allowing rotation at locations where bending moments would otherwise be high. A typical example is converting a rigid frame into a pinned frame, adding internal hinges to arches, or turning a fixed beam into a simply supported beam.

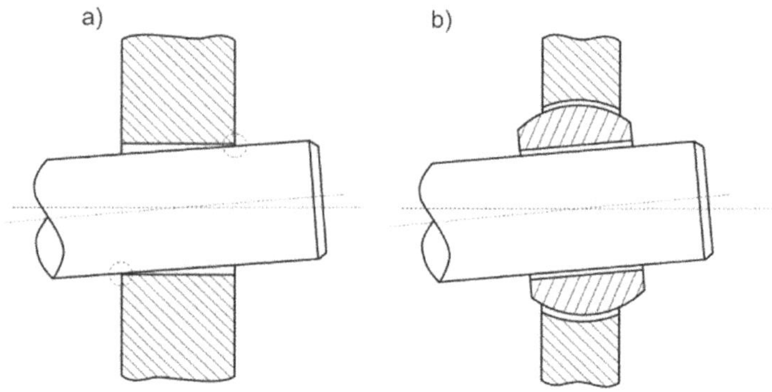

FIGURE 7.3 Deliberate weakness adding extra degrees of freedom.

7.4.3 Energy-Absorbing Deliberate Weaknesses

The mechanism behind this category of deliberate weaknesses is that they are characterized by a large amount of absorbed energy. As a result, the maximum load on the structure is reduced. Crumple zones in road cars and crash cones in racing cars are common examples. In case of impact, the deliberate weaknesses deform and reduce the impact load on driver and passengers.

Buffer systems in railway is one more example. They incorporate elastic elements whose purpose is to absorb impact energy in a controlled way during a collision, to reduce damage.

Modern buildings in seismic zones include components designed to deform or fail first (e.g., steel shear links). These are sacrificial elements that act as seismic fuses and have a similar purpose. In the event of earthquake, the weakened beams undergo plastic deformation which consumes a large amount of earthquake energy. As a result, the seismic fuses absorb seismic energy, localize damage, and prevent failure of critical load-bearing members. The seismic forces on the structure are significantly diminished and the likelihood of structural collapse during an earthquake is reduced significantly. The seismic fuses can be replaced after an earthquake, keeping the overall structure intact and reducing repair costs and casualties.

Deliberate weaknesses can also be found naturally. Thus, the energy-dissipating properties of the deliberate weaknesses known as 'sacrificial bonds' account for the formidable fracture toughness of natural materials. The high toughness of natural materials is due to the reversible, molecular-scale energy-dissipation mechanism provided by sacrificial bonds in the structural molecules (Fantner et al., 2006). These tough natural materials have been mimicked in the fabrication of synthetic polymeric materials where part of the synthesis is the introduction of energy-dissipating sacrificial bonds.

7.4.4 Case Study: Deliberate Weaknesses Incorporating an Elastic Element

Elements with reduced elastic constants are other examples of energy-absorbing deliberate weaknesses. In case of impact, part of the kinetic energy E_k of the impacting object is converted into elastic strain energy U. The elastic strain energy U is equal to the work done by the maximum force P during the impact:

$$U = \frac{1}{2} P\delta \tag{7.1}$$

where δ is the maximum displacement (in the elastic region) due to the impact. Reducing the elastic constant increases the displacement δ which, according to equation (7.1), leads to a reduced impact force P.

The technique of reducing the maximum stress by incorporating an elastic element as a deliberate weakness is a technique that has been used to reduce the maximum stress in a cable subjected to large inertia forces due to sudden stopping.

Consider an object (1) with mass m, supported by a cable (3) with cross section area A, wound on a rotating hoist drum (2) (Figure 7.4a). The weight is moving downward with a particular speed v when the rotation of the drum is suddenly stopped (for example, the drum is suddenly locked).

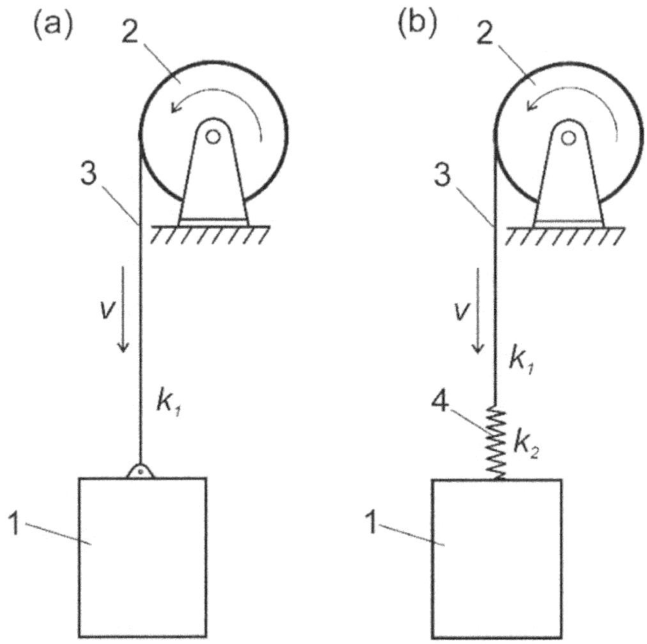

FIGURE 7.4 Hoist drum (a) without and (b) with a deliberate weakness.

Suppose that a deliberate weak link has been introduced in the form of an additional component '4'. This component effectively acts as a spring connected in series (Figure 7.4b), whose elastic constant k_2 is significantly smaller than the elastic constant k_1 of the cable '3'. The equivalent stiffness k_e of both the cable and deliberate weakness (Figure 7.4b) is given by the well-known relationship (Samuel and Weir, 1999):

$$k_e = \frac{1}{\frac{1}{k_1} + \frac{1}{k_2}} = \frac{k_1 k_2}{k_1 + k_2} = \frac{k_2}{1 + k_2 / k_1} \tag{7.2}$$

For $k_2 / k_1 << 1$, $k_e \approx k_2$.

The strain energy U accumulated by elastic elements with effective constant k_e is given by $U = \frac{P^2}{2k_e}$ (Gere and Timoshenko, 1999) where P is the maximum force acting on the elements.

Consider the body '1' with mass m moving with velocity v due to unwinding the cable '3' from the drum. Suppose that at some instant the drum suddenly stops. Upon stopping, a fraction γ of the kinetic energy $E_k = \frac{mv^2}{2}$ of the body is transformed entirely into strain energy of the assembly. The dynamic force P resulting from the sudden stop can then be evaluated by equating U and γE_k:

$$\frac{P^2}{2k_e} = \gamma E_k \; : P = \sqrt{2k_e \gamma E_k} \tag{7.3}$$

From the last equation, it is clear that reducing the equivalent elastic constant from k_1 to k_e by introducing an elastic element in series results in a significant decrease of the magnitude of the dynamic force.

This is an example demonstrating how combining domain-independent knowledge of energy-absorbing deliberate weaknesses and domain-specific knowledge from stress analysis yields a significant reduction of the risk of overstress failure.

7.5 DELIBERATE WEAKNESSES PREVENTING EXCESSIVE DAMAGE

7.5.1 Deliberate Weaknesses Limiting the Spread of Damage

The central ideas behind this category of deliberate weaknesses are (i) to restrict the spread of damage and (ii) to prevent excessive damage accumulation by decoupling the damaged parts of the system from the rest of the system.

An example of this type of deliberate weakness is the sectioning of a pipeline with weak interfaces between the individual sections. As a result, the propagation of a crack appearing in one of the sections is arrested at the weak interface and the crack is confined within a single section only. Consequently, the cost of repair is reduced to the replacement of a single section only. Without restricting the spread of damage, a crack appearing in one section could travel along a very large distance of the pipeline (particularly cleavage cracks in steel pipelines at sub-zero temperatures).

Deliberate weaknesses of this type can even increase the load-carrying capacity and fault tolerance of monolithic columns by segmenting them and creating weak interfaces between the individual segments. As a result, damage is confined in small area and its propagation is severely inhibited.

Connecting modules with deliberately weakened links is an important protection technique. In case of overstress, the weak links fail, and the excessive stress is not transmitted to the expensive modules. An example of this type of deliberate weaknesses is a network with hubs connected with deliberate weak links. Suppose that the links are sufficiently strong to provide the necessary connections and sufficiently weak to fail in the event of overstress and protect the rest of the network from damage escalation. Then, the result is damage contained in a small region while the rest of the network is not affected.

Compared to some traditional protection methods which require significant investment such as condition monitoring or strengthening the links in the network, a technique based on deliberate weak links is a low-cost solution constantly ready to operate.

7.5.2 Deliberate Weaknesses Protecting from Excessive Levels of Damage Accumulation

The underlying mechanism for this category of deliberate weaknesses is protection against excessive damage accumulation by making damage accumulate in inexpensive sacrificial links instead of accumulating in the valuable parts of the system.

Familiar examples of deliberate weaknesses from this category are the sacrificial anodes attached to underground pipes, ship hulls, storage tanks, etc. By corroding preferentially, they save the metal surface to which they are attached. Sacrificial coatings such as zinc coatings are similar in function and are commonly used for corrosion protection of outdoor structures.

Rubber coatings covering the segments of conveyor belts perform a similar function. Damage accumulates in the inexpensive rubber coating which, after certain level of damage, is replaced and the expensive segments of the conveyor belt are protected.

Inserts into spray nozzles is another example. By undergoing wear preferentially, the inserts act as deliberate weaknesses where damage accumulates, thereby protecting the expensive spray nozzle from excessive wear.

The wear piece behind the impeller of a sludge pump also carries out the function of a sacrificial wear element. The wear piece takes the brunt of heavy wear and protects the expensive pump.

7.5.3 Deliberate Weaknesses Providing Stress Relaxation

Common examples of deliberate weaknesses of this category are the expansion offsets accommodating dilatations from thermal expansion/contraction of pipes, railways, bridges, girders, etc., which could otherwise destroy the structures.

The deliberate gaps introduced at regular intervals on a pavement have a similar purpose.

Designing joints free to rotate in trusses (instead of welding) eliminates bending and reduces the magnitude of the internal stresses.

Segmenting structures by providing surfaces free of shear stresses also reduces the magnitude of the internal stresses.

Another example involves aircraft *fuse plugs* in landing gear tires. Their purpose is to prevent dangerous tire explosions caused by overheating during heavy braking, or high-energy landings. An overheated tire blowout can hurl shrapnel-like debris at lethal speeds. To mitigate this, the wheel rim is fitted with small plugs made from a fusible alloy, often a eutectic metal mixture with a low melting point (about 177–204°C), housed in a steel body threaded into the rim. Under normal conditions, the plug seals the pressurized air or nitrogen inside the tire.

If brake heat is conducted into the wheel after intense braking, the fusible alloy melts at its calibrated temperature, well before the tire approaches a dangerous burst pressure. The melted alloy opens a vent, allowing the tire to deflate in a controlled manner rather than explosively.

7.5.4 Deliberate Weaknesses Acting as Frangible Objects

Frangible objects/structures are deliberately designed to break easily upon overstress. The purpose is to protect the more valuable object and minimize damage. The lighting towers in airports is a typical example. Upon collision with an aircraft, they easily break and protect the aircraft from excessive damage. Frangible light poles are another example. They fail easily upon collision with a car, thus reducing the consequences to drivers and passengers.

Frangible roof panels on storage tanks are also examples of frangible objects. The roof panels are designed to blow off safely during overpressure. The purpose is to prevent catastrophic tank rupture and direct the explosion away from personnel.

Weakly fixed objects are designed with a similar purpose. For example, the detachment of an aircraft engine upon emergency landing protects the aircraft from fire.

Safety rings or tamper-evident caps on medicine bottles that snap when first opened are also examples of frangible objects. Their purpose is to indicate tampering and reduce risk of unnoticed contamination.

7.6 DELIBERATE WEAKNESSES DEFLECTING DAMAGE FROM PLACES WHERE THE COST OF FAILURE IS HIGH

The mechanism underlying this category of deliberate weaknesses consists of deflecting the location of damage accumulation and failure away from parts where the cost of failure or cost of intervention is high.

The example given in Figure 7.5 depicts a pipe (1) carrying fluid with debris. The debris tend to clog the pipe and if this happens in zone '3', with difficult access, a costly and lengthy intervention will be required.

To avoid this, a deliberate weakness deflecting the clogging has been installed in an easily accessible place in the form of a deliberately narrowed section with ribs '2' which makes clogging in this section easier compared to the rest of the pipe. This arrangement permits the deliberate weakness to be removed, cleaned, and installed back in place. Without the deliberate weakness, clogging could occur in zone '3' which requires costly intervention associated with significant delays.

The deliberately weak interfaces in layered structures serve also as deliberate weaknesses deflecting damage. Propagating cracks are deflected along the weak interfaces and their penetration into the components is stopped or delayed significantly (Figure 7.6).

Narrow, deliberately made pre-cracks along underground cables, are extended if the temperature drops significantly below zero and causes the ground to freeze (Figure 7.7). As a result, the formation of cracks across the cables and severing of the cables is prevented (Altshuller, 1999).

The application of this category of deliberate weaknesses transcends engineering. Thus, in computer security, 'sacrificial hosts' built outside the internet firewall of an organization are intended to lure and bait potential attackers, delay them, issue warning, and even obtain information about the identity of the attackers.

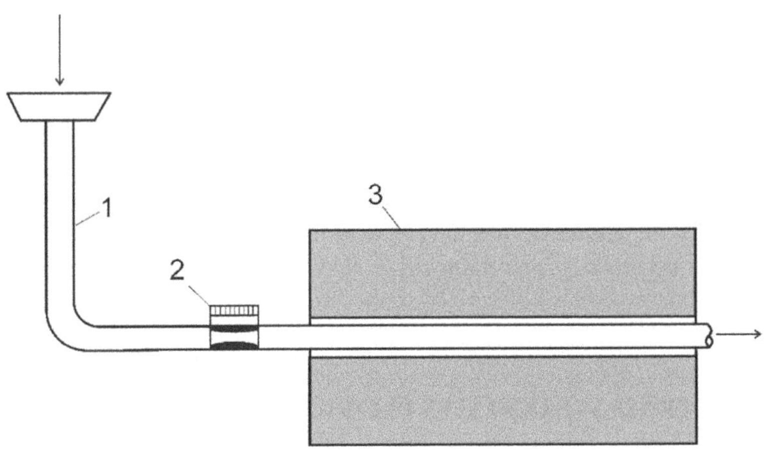

FIGURE 7.5 Reducing the risk of long delay caused by blockage by deflecting the potential failure to a place where the intervention for repair is easier.

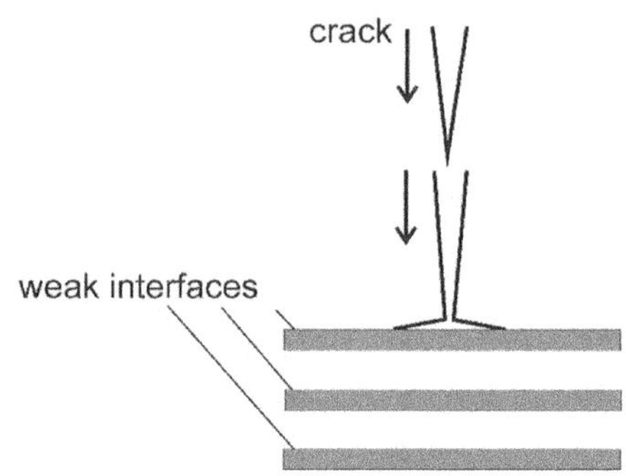

FIGURE 7.6 Weak interfaces delaying a propagating crack.

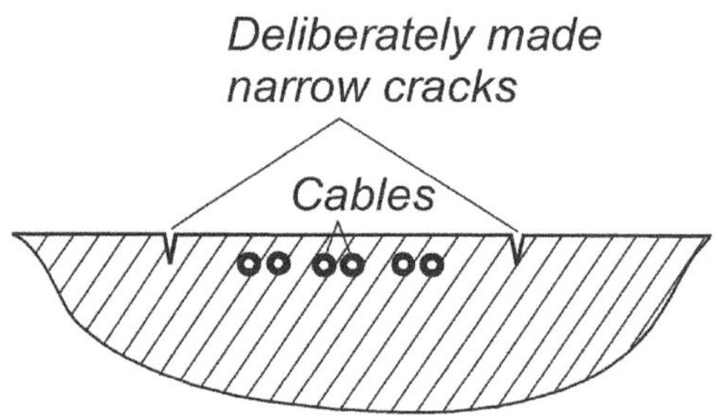

FIGURE 7.7 Protecting underground cable lines from cracks caused by freezing of the ground (Altshuller, 1999).

Pilot projects in project management have a similar function. A relatively small investment is allocated for the pilot project and if failure does occur, the loss is relatively small because it is confined within the budget of the pilot project not within the budget of the expensive main project. In addition, if failure does occur in the pilot project, valuable lessons are still learned about the technical feasibility of the product, the marketplace, and customers. As a result, the risk of a significant loss of investment is reduced significantly.

Leaving a deliberate weakness to be attacked in order to regroup forces and concentrate on striking elsewhere has always been an important element of the military tactics. The well-known 'pincer manoeuvre' leaves a weakened centre to be attacked by advancing enemy forces followed by ambushing the flanks of the enemy formation by concealed side forces.

7.7 DELIBERATE WEAKNESSES PROVIDING WARNING

This category of deliberate weaknesses provide a warning about increasing levels of stress/strain which could damage components and structures. Mechanical fuses on cables are a good example of this type of deliberate weaknesses (US patent US2011/0027007A1).

Figure 7.8 depicts a retrofittable cable fuse, where '1' is the load-carrying cable, '2' are clamps through which the mechanical fuse '3' is attached to the cable. In normal operating conditions, the load is carried by the deliberately weakened fuse (part 3 in Figure 7.6). Should load exceed a dangerous limit, the fuse 3 deforms and the load is transferred to cable 1.

Sand probes in pipelines are another example. They provide warning about excessive erosion due to a large amount of sand particles in pipelines transporting crude oil.

In the area of security, tapes which easily delaminate and fracture, are used to provide warning against tampering or unauthorized access.

7.8 DELIBERATE WEAKNESSES PROVIDING RAPID ACCESS

In many situations, rapid access to emergency resources is crucial. This applies to medical supplies, alarm push-buttons, ammunition, and similar items. In such cases, the deliberate weakening of a casing or housing is employed to ensure quick access during an emergency. A common example is the alarm push-button placed behind a breakable glass cover, which represents a purposeful weakness in design.

Other examples include fuse holders with quick-release tabs that allow tool-free replacement, as well as weakened access plugs designed to be drilled or knocked out easily to open pipes.

7.9 DELIBERATE WEAKNESSES PROVIDING QUICK ESCAPE

The idea is to provide quick escape in critical circumstances by deliberately weakening components in the system. Such are the various quick-release pins, break-away escape links, break-away buckles for pet-animal collars, and break-away links attaching lifeboats. The break-away tagline used in air rescue operations includes such a deliberate weakness. In the event of entanglement of the tagline, the break-away weak link allows the tagline to be easily broken by the personnel thereby allowing the helicopter to escape.

This category of deliberate weaknesses is also found in nature. Some animals lose a particular body part as a defence mechanism (*autotomy*). The gecko lizard is a

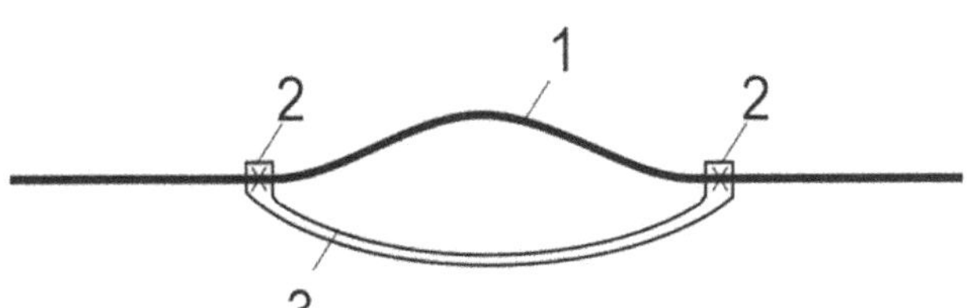

FIGURE 7.8 Retrofittable cable mechanical fuse (US patent US20110027007A1).

well-known example. The gecko's tail contains a deliberately weak connective tissue that can easily break when the animal is in danger.

7.10 DELIBERATE WEAKNESSES TRIGGERING PROTECTION SYSTEMS

The purpose of this category of deliberate weaknesses is to provide indirect protection by triggering a specially designed protection system in case of overload.

Common examples are the deliberately weakened plugs in fire protection systems consisting of an alloy with low melting point or a weakened glass bulb. In case of fire, the deliberately weakened plug fails, and the sprinkler fire protection system is activated.

The hydraulic damper secured by weakened pins is another example. In case of strong earthquake forces, the pins shear and release the hydraulic damper which mitigates the damage to the building.

Honeytokens in cybersecurity offer another example. These are fake user credentials or data intentionally planted within a system and are never meant to be accessed by legitimate users. If an intruder attempts to use or access a honeytoken, it signals a potential breach. This action can then trigger alerts, activate defensive mechanisms, or log the activity for further investigation, enabling a swift and targeted incident response.

7.11 THE EQUAL-RELIABILITY/EQUAL-STRENGTH CONCEPT AND THE DELIBERATE WEAKNESSES CONCEPT

A discussion on deliberate weaknesses in design will be incomplete if the relationship between the deliberate weakness concept and the equal-reliability concept is not clarified.

A common reliability allocation strategy in engineering design is to make all parts with comparable reliability or strength and not to leave weak links. Such is, for example, the AGREE methods described in (Ebeling, 1997). The essence of this strategy is to allocate to each module in a system an equal share of the reliability of the system.

In Thompson (1999), for example, it has been advocated that the safety margins of all components are selected in such a way that a maximum 'distance' to the constraints preventing the separate failure modes from occurring is maintained. The underlying idea is that this strategy will maximize system reliability and offer maximum protection against uncertain variations caused by loads and environmental conditions.

Another reliability allocation strategy (Ebeling, 1997) is to optimize the reliability allocation, so that the cost of achieving the required reliability is minimized.

None of the described reliability allocation strategies considers the cost of component failure. Following any of these reliability allocation strategies could be justified only if the cost of system failure due to failure of any component/part is approximately the same. In reality, the cost of system failure due to failure of the individual components varies significantly. Neither designing the parts in the system with equal (comparable) reliability nor achieving the system reliability in the least-cost manner minimizes the expected loss given that failure occurs.

The expected loss given failure $\bar{C}_f$ of a system with M components logically arranged in series is given by (Todinov, 2004)

$$\bar{C}_f = \sum_{k=1}^{M} \frac{\lambda_k}{\lambda_1 + \ldots + \lambda_M} \bar{C}_{k|f} \tag{7.4}$$

where λ_i, $i = 1,\ldots,n$ are the hazard rates characterizing the separate components, and $\bar{C}_{k|f}$ is the expected loss given failure, characterizing the ith component.

Consider a simple system consisting of two blocks only (A and B). The cost of repair C_A of block A is much higher than the cost of repair C_B of block B ($C_A \gg C_B$). Suppose that λ_1 is the failure rate of block A while λ_2 is the failure rate of block B. For the sake of simplicity, let $\lambda_1 \approx \lambda_2$. According to equation (7.4), the expected loss given failure $\bar{C}_f$ is given by

$$\bar{C}_f = \frac{\lambda_1}{\lambda_1 + \lambda_2} C_A + \frac{\lambda_2}{\lambda_1 + \lambda_2} C_B \approx (1/2)C_A + (1/2)C_B \tag{7.5}$$

Introducing a deliberate weakness in block B will make its hazard rate λ_2' much greater than the hazard rate of block A: $\lambda_1 / \lambda_2' \approx 0$. The expected loss given failure now becomes:

$$\bar{C}'_f = \frac{\lambda_1}{\lambda_1 + \lambda_2'} C_A + \frac{\lambda_2'}{\lambda_1 + \lambda_2'} C_B = \frac{1}{1 + \lambda_2' / \lambda_1} C_A + \frac{1}{\lambda_1 / \lambda_2' + 1} C_B \approx C_B \tag{7.6}$$

As a result, the consequences from failure will be reduced because it is much more likely that block B will fail in case of overload. The expected cost of repair if a deliberate weakness is present is significantly smaller compared to the expected cost of repair if no deliberate weakness is present.

Creating components with comparable reliability or strength, for example, will make the failure occurrence to be equally likely in any component, irrespective of its cost of failure, which, in many cases is highly undesirable. In a number of cases, to reduce the cost of failure, it is of critical importance that failure occurs in a controlled manner and at a predictable location.

Consider, for example, sections of subsea oil and gas production equipment. Some of them are located on a floating platform and others are located on the seabed. The design should be made in such a way that if failure occurs, the likely failure location is in the sections on the floating platform. On the floating platform, the cost of intervention for repair is much smaller compared to the cost of intervention for repair in the subsea sections. The intervention for repair in the subsea sections is a complex process which typically requires remotely operated underwater vehicles or oil rigs.

8 Improving Reliability and Reducing Risk by Inversion

8.1 THE METHOD OF INVERSION

The underlying idea of the method of inversion is to avoid failure modes by inverting relative position, orientation, motion, functions, features, properties, mode of thinking, or by introducing inverse states.

A failure mode that appears in a specific position, orientation, motion, or state often vanishes when the system or parameter is shifted to the opposite state while maintaining the system's required functions.

The method of inversion also achieves risk reduction by providing a different frame of thinking which helps to see easily the risk reducing solution.

The inverse strategy for reliability improvement and risk reduction, which is the focus of this chapter, has always been an important technique in problem solving. It was already known to Pappus of Alexandria about 320 AD, who described it as 'beginning with the desired outcome and working backwards until something known is reached'. The famous algebraist *Carl Gustav Jacobi*, whose favourite maxim was '*invert, always invert*', used the inverse strategy with great success to solve many hard mathematical problems. An important example of the inverse strategy in problem solving is the recursion, widely used for solving difficult combinatorial problems (Wirth, 1976). A typical approach in recursion is reducing the initial problem of size n to two or more problems of size $n-1$, which in turn are reduced to problems of size $n-2$ and so on, until problems with trivial solutions are obtained. Starting from the trivial solutions and moving backwards helps assemble the solution of the original problem.

It is then no surprise that inverse strategy features in texts related to creativity and inventive problem solving. The inverse strategy, for example, is at the heart of many inventions and Altshuller's TRIZ framework (Altshuller, 1984, 1996, 1999) identified a number of useful applications of this technique.

The inverse strategy is often used in planning projects, operations, or processes where success is critically dependent on whether people, equipment, and resources are available when and where needed. This is particularly relevant to cases where critical deadlines need to be met. In order to reduce the risk of failure, the planning of the project/process is often started from the desired outcome and progressed backwards. Starting from the desired end result and planning backwards helps to determine what

DOI: 10.1201/9781003679301-8

resources are necessary to achieve the goal and to ensure how to make the necessary resources available at the right place and at the right time.

Despite the progress made in using the inverse strategy for problem solving and planning, no treatment currently exists on the application of the inverse strategy for reliability improvement and risk reduction. Isolated applications of reliability improvement and risk reduction based on inversion do exist, but these applications have not been linked with the generic method of inversion.

The inverse strategy is a powerful method for improving reliability and reducing risk and will be referred to as *the method of inversion* (Todinov, 2017c).

In what follows, the different techniques and mechanisms through which the method of inversion reduces risk are discussed in detail.

The classification of the techniques for improving reliability and reducing risk based on the method of inversion, discussed in this chapter, is given in Figure 8.1. This classification introduces a structure in implementing the method of inversion and simplifies the process of searching for an appropriate solution. In addition, the structure provided by the classification makes the application of the method of inversion less dependent on inspiration and lateral thinking leaps. The classification is easy to use and does not require a significant investment of time. In addition, the classification acts as a useful checklist which helps avoid the omission of a relevant technique based on the method of inversion.

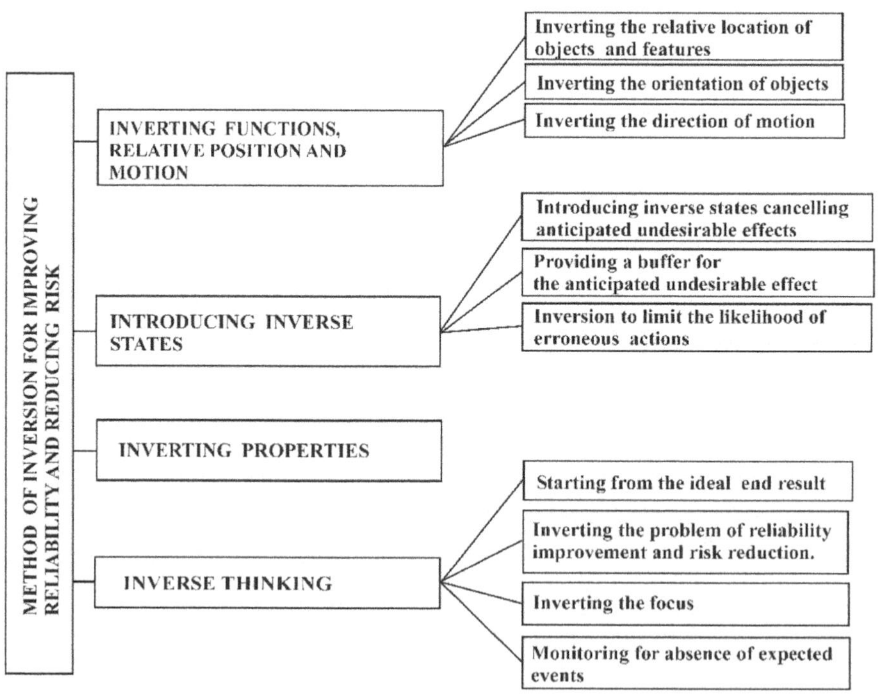

FIGURE 8.1 The method of inversion for improving reliability and reducing risk.

8.2 IMPROVING RELIABILITY BY INVERTING FUNCTIONS, RELATIVE POSITION, AND MOTION

Most of the processes in engineering are reversible. The basic function delivered by a system often does not change if the relative position, motion, or features of interacting objects are reversed. *By reversing functions, relative position, motion, or features, the basic required functions are often preserved whilst critical failure modes disappear or are mitigated.* This is an important idea behind the method of inversion. Inverting motion, relative position of objects and the location of features often eliminates failure modes, simplifies the interaction and results in enhanced performance. An important feature of the method of inversion is that, by applying this method, often the reliability of a product is improved at no extra cost.

The method of inversion works because between two opposite states of a system parameter (A and its inverse $\bar{A}$) one of them will usually be with maximized resistance against a particular critical failure mode and superior overall reliability.

An example of eliminating failure modes by inverting a function can be given with the fail-safe air brakes of trucks. Instead of air pressure energizing the breaks when these are needed, the function of air pressure is inverted. The air pressure keeps the brakes released which permits the truck to move. In the case of a low pressure in the air line due to a puncture or failure of the compressor, the brakes are applied securing the truck. In this way, a dangerous failure mode is avoided: loss of air pressure and inability to apply breaks when needed.

A simple illustration of the method of inversion can also be given with making a bolt connection with threaded blind holes. Inverting upside down the orientation of the threaded blind holes prevents the collection of dirt, abrasive particles, and moisture in the threaded holes, because gravity helps to clean the holes. The result is reduced corrosion and wear and extended life of the bolt connection. The method of inversion works because the inverse orientation of the blind hole is more resistant to the collection of dirt, abrasive particles, and moisture. A very similar situation is present in drilling vertical blind holes in components by a robot, on a manufacturing line. This operation is associated with the failure mode: 'metal chips stuck at the bottom of the blind hole because of gravity'. However, if the hole is drilled on a component positioned upside-down, the failure mode is eliminated because gravity now helps to clean the hole.

Often, making a moving object stationary while moving the stationary objects in the opposite direction (inverting the direction of motion) eliminates major failure modes and results in a significantly improved performance and risk reduction. This idea underlies the Cosworth® sand-casting process (Campbel, 2015) where the molten metal is never poured down into the sand mould as is the case in the classical sand-casting process. The molten metal flows in the opposite direction (uphill) into the mould. This type of metal flow eliminates turbulence and mitigates a major failure mode: 'trapping oxides into the metal,' which reduces significantly the fatigue strength of the cast component.

Moving parts exhibit more failures compared with stationary parts. This is usually due to increased kinetic energy, wear, fatigue, vibration, heat generation, and erosion associated with moving parts.

Reducing the number of moving parts is an efficient way of improving the reliability of a system and inverting the motion can be used to reduce the number of moving parts.

Suppose that a required function is guaranteed by the relative motion of three components with respect to each other. If an extra uniform motion with opposite direction to the motion of the first component is imparted to all three components, the first component will become stationary and the motion of the rest of the components will become a composition of their motion before the inversion and the motion of the first component taken with a negative sign. The relative motion of the components will remain the same, but the reliability of the system will be improved because the number of moving parts will have been decreased by one.

There are numerous cases where inverting the relative location of features eliminates failure modes, enhances reliability, and reduces risk.

A historic example is the flanged wheel of a railroad train. Flanged wheels are needed to prevent the train from sliding off the track. Initially, the railroad tracks were flanged instead of the wheels and thousands of miles of unsafe railroad tracks were manufactured, with thousands of miles of unnecessary flanges. The reliability improved dramatically when inversion was applied to the location of the flange. The location of the flange was shifted from the railway track to the wheel. In this way, critical failure modes were eliminated while preserving the basic function of keeping the train from sliding off the track. As a result, the railway safety improved significantly.

The inversion method will be illustrated by a case study involving an alarm circuit.

8.2.1 Case Study: Improving the Reliability of a Switch

When a variable force F is applied during the operation of the normally open switch shown in Figure 8.2a, the soldered joints on the printed circuit board are subjected to repeated high-magnitude pulsating stress. This occurs because the excessive variable force applied to the button is directly transferred as a variable load to the soldered points. As a result, a fatigue crack forms at the soldered points connecting the switch (Figure 8.2a), often well before the expected service life is reached.

By applying inversion which transforms the switch from normally open to normally closed, the activation mechanism changes. Instead of closing normally open contacts (Figure 8.2a), activation now involves opening normally closed contacts (Figure 8.2b).

As a result, an excessive force applied to the push-button no longer causes an increased stress in the soldered points on the circuit board. Due to inversion, the detrimental fatigue loading on the soldered points is diminished significantly, which diminishes significantly the likelihood of premature fatigue failure. Leveraging the asymmetric response from inversion effectively removes a crucial failure mode.

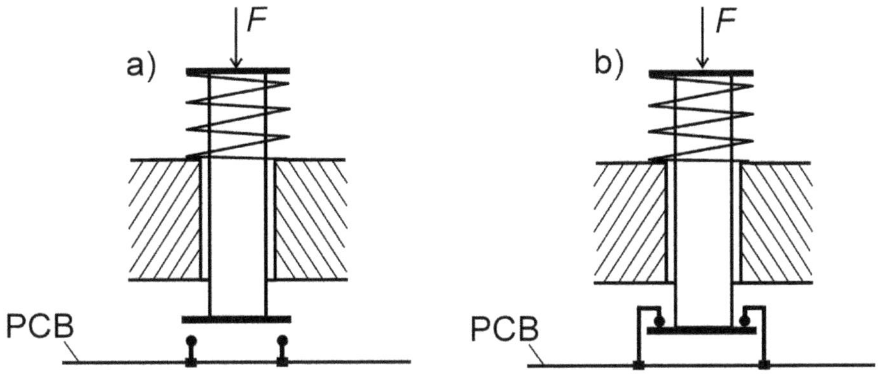

FIGURE 8.2 (a) A normally open mechanical switch whose repeated actuation causes fatigue failure of the soldered points on the printed circuit board; (b) enhancing the reliability of the mechanical switch by inversion from normally open to normally closed contacts.

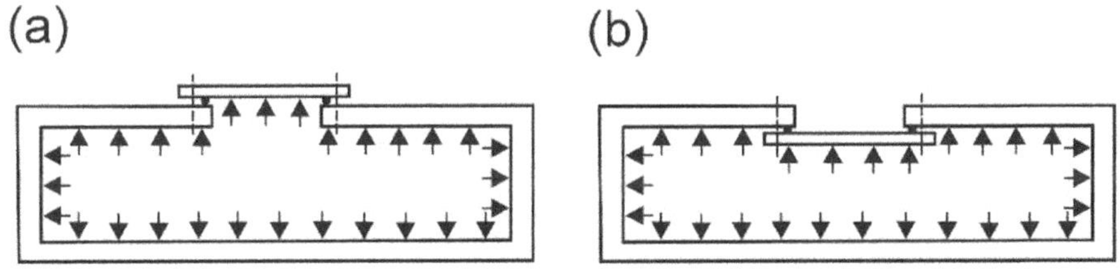

FIGURE 8.3 (a) Improving the reliability of a container under pressure by (b) inverting the relative position of the cover.

8.2.2 Improving Reliability by Inverting the Relative Position of Objects

The next major mechanism through which inversion improves reliability is that *inversion of relative position often increases the response of the component/system to loading stresses.*

Inversion can be used to achieve a self-reinforcing response. Such an application is given in Figure 8.3, in the design of a cover for containers under pressure.

For the design in Figure 8.3a, the loading stresses in the screws fixing the cover can be reduced and the reliability of the seal can be increased significantly by *inverting the position of the cover with respect to the pressure vessel.* The design from Figure 8.3b channels the force induced by the internal pressure into a self-reinforcing response which helps to form a reliable seal and prevent leakage. Increasing pressure increases the magnitude of the self-reinforcing response, strengthens the seal, and increases the resistance to increased pressure. In this example, reliability has also been improved at no extra cost by applying the method of inversion.

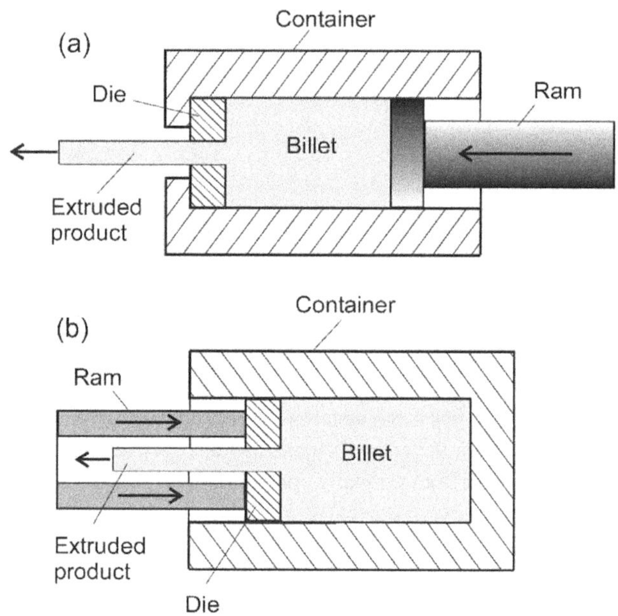

FIGURE 8.4 (a) Direct and (b) indirect extrusion.

8.2.3 Case Study: Inverting Motion to Reduce the Wear of an Extrusion Container

The extrusion process depicted in Figure 8.4a suffers a failure mode: friction between the billet and the extrusion container causes intensive wear of the container which diminishes its life. By using inversion, this failure mode can be eliminated and the wear of the container significantly reduced.

The solution, by using inversion of motion (Figure 8.4b), is to move the die in the opposite direction, which makes the billet material stationary. As a result, friction between the billet with the walls of the container no longer exists and the wear of the container is eliminated.

8.3 IMPROVING RELIABILITY BY INVERTING PROPERTIES

Inverting material properties to enhance reliability involves modifying, selecting, or designing materials to eliminate or reduce the likelihood of expected failure modes. Consider two components, *A* and *B*, that operate in contact, and component *A* is significantly more expensive than *B*. If *A* is the softer material, wear damage will accumulate on the more expensive component, leading to higher maintenance or replacement costs.

By inverting their mechanical properties, specifically, making component *B* softer than component *A*, the functional requirements remain unchanged, but the wear shifts to the less expensive component *B*. This strategy protects the more valuable component and reduces overall lifecycle costs, as component *B* can be replaced more economically.

A practical example of this strategy is found in vehicle braking systems. Brake pads are deliberately made softer than brake discs, which ensures effective braking while directing wear to the cheaper brake pads rather than the significantly more expensive discs.

If eliminating a failure mode requires dissipating heat from a material with low thermal conductivity, a material with high thermal conductivity is used instead.

Conversely, in turbine blades, for example, ceramic coatings with low thermal conductivity are applied to metal surfaces. These coatings help keep the underlying metal cool and prevent oxidation and creep.

Another well-known example from flanges bolted on pressure vessels involves reducing the range of pulsating force in the bolts to improve their fatigue life. This is achieved by inverting the elastic constants of the bolts and the flange while maintaining the required clamping force. Specifically, the bolts' high elastic constant is reduced, and the flange's elastic constant is increased. This invertion of properties preserves the necessary clamping force but reduces the amplitude of force variation in the bolts, resulting in significantly improved fatigue life and greater reliability.

8.3.1 Case Study: Improving the Maintenance of a Glass Filter

Figure 8.5a illustrates a glass filter used in a production system that exhibits a critical failure mode: cleaning the filter is both difficult and time-consuming. The *availability* of the production process is defined as:

Availability = Uptime / (Uptime + Downtime)

where *Uptime* refers to the period during which the system operates normally, and *Downtime* denotes the time required for repair or maintenance.

The lengthy cleaning process of the glass filter significantly increases downtime, thereby reducing the overall availability of the system. However, by applying *inversion of properties*, the maintainability of the glass filter can be enhanced, leading to a substantial improvement in system's availability.

The solution (Altshuller, 1996; Figure 8.5b) is achieved by inverting the material properties: solid glass regions are transformed into voids, and voids are replaced with solid glass. This is accomplished by packing multiple solid glass rods into the filter housing. As a result, the failure mode associated with slow filter cleaning is eliminated, leading to improved maintenance efficiency and production availability.

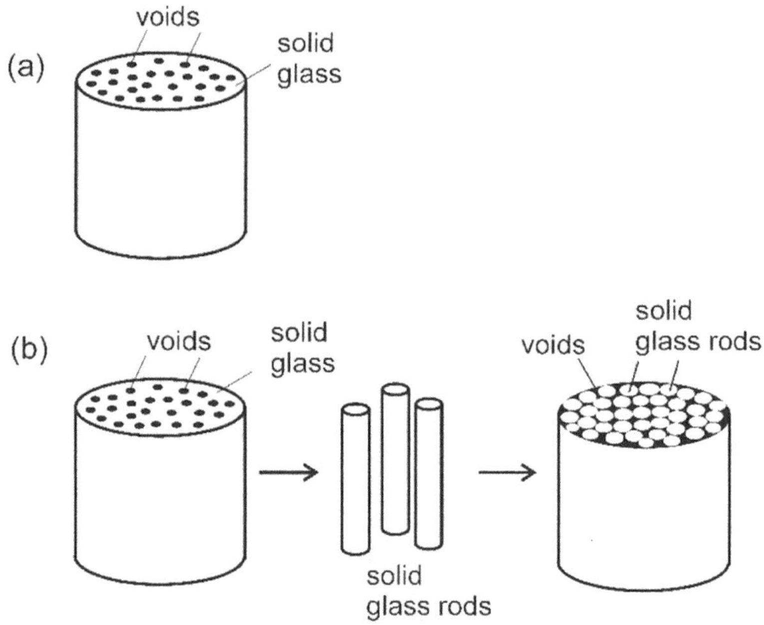

FIGURE 8.5 (a) A glass filter that is difficult to clean; (b) improved maintenance efficiency of a glass filter, obtained by inverting properties.

8.4 IMPROVING RELIABILITY AND REDUCING RISK BY INTRODUCING INVERSE STATES

8.4.1 Inverse States Cancelling Anticipated Undesirable Effects

An intentional state, opposite to the anticipated undesired state, can be introduced to cancel or significantly reduce the undesirable effect. When the two states are superimposed, the result is either elimination or substantial attenuation of the unwanted state.

In acoustics, this principle is employed in noise-cancelling headphones, which reduce the risk of hearing damage caused by excessive noise. A sound wave with the same amplitude but opposite phase is emitted to counteract incoming noise, resulting in a significant noise reduction and improved auditory safety.

This same principle forms the basis of active vibration control. The system typically includes vibration sensors (such as accelerometers), controllers, and actuators. The sensors detect unwanted vibrations and send signals to the controller, which then directs the actuators to generate counter-vibrations. These cancel the original vibrations through destructive interference. Advances in sensor, actuator, and computing technologies have made active control methods increasingly cost-effective and widely accessible.

Deliberately introduced inverse states that cancel undesirable anticipated effects are central to many temperature-compensation circuits, which are designed to mitigate the impact of heat on the performance of electronic devices. For instance, a Wheatstone bridge can generate an inverse response to temperature changes, effectively cancelling out thermal effects. This principle underpins numerous precision measurement techniques known for their minimal error.

In mechanical engineering, a classic example is the compensation pendulum in clocks. Here, the thermal expansion of the pendulum rod is counteracted by components (also part of the pendulum rod) that expand in the opposite direction, ensuring that the length of the pendulum rod and the period of oscillation remain constant despite temperature fluctuations.

Another example of an inverse state as a counterbalancing force is the use of counterweights in tower cranes. These enhance the crane's balance and stability. Similarly, gate fail-safe valves utilize hydraulic pressure to oppose a compression spring. During normal operation, the hydraulic pressure keeps the valve open. If the hydraulic system fails, the spring expands, automatically closing the valve and returning it to a safe position.

8.4.2 Case Study: Improving the Precision of a Measuring Device

The length of the measuring rod '1' in the measuring device shown in Figure 8.6 changes with temperature due to thermal expansion and contraction (Figure 8.6a). In a high-precision instrument, even slight dimensional variations can lead to significant measurement errors. To ensure accuracy, it is essential to minimize distortion in the measured values caused by thermal variations.

Although alloys with very low coefficients of thermal expansion, such as Invar, can reduce this variation, they still exhibit measurable expansion or contraction and may not provide an acceptable solution. By introducing an inverse state, a solution can be found that cancels the thermal expansion of the metal probe.

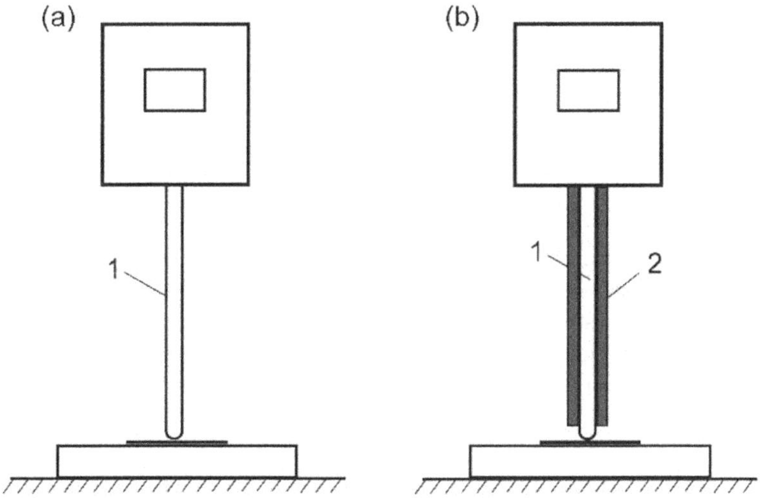

FIGURE 8.6 (a) A measuring device with dimensional variations of the measuring rod due to thermal variations; (b) a measuring device with reduced dimensional variations due to thermal variations.

A compensating inverse state for the thermal expansion of the measuring probe can be achieved by introducing a shrink-fitted sleeve '2' (Figure 8.6b). The sleeve applies radial compression to the rod, causing elastic axial elongation due to Poisson's effect (US Patent 3528206A). As the temperature increases, the sleeve expands thermally, reducing the radial compression. This reduction leads to elastic axial contraction of the rod, thereby compensating for thermal expansion.

8.4.3 Inverse States Buffering Anticipated Undesirable Effects

Inverse states are often employed to buffer anticipated conditions that may have undesirable effects. This approach is fundamental to reducing the risk of failure in heat-generating zones. Components in close contact, such as piston and cylinder assemblies, generate heat through relative motion. If not properly dissipated, this heat can lead to severe wear, reduced strength, and structural deformation. The risk of failure can be significantly reduced by introducing an inverse state, such as cooling one of the components (e.g., the cylinder), to dissipate heat. This intentional cooling lowers friction and wear, thereby enhancing system reliability.

Cold expansion, commonly used in the aviation industry to induce compressive stresses around fastener holes, exemplifies an inverse state that counteracts anticipated detrimental effects (Fu et al., 2015). The process involves driving a tapered mandrel through the fastener hole, plastically expanding it. This creates a compressive residual stress field around the hole, which counteracts operational tensile stresses. As a result, the initiation and propagation of fatigue cracks at the hole edge are significantly impeded.

To counteract surface tensile stresses from loading and enhance fatigue resistance, shot peening, a process that induces compressive residual stresses at the surface, has become a critical element of manufacturing technology (Niku-Lari, 1981; Bird and Saynor, 1984; Burrell, 1985). In components such as compression springs, shot peening generates compressive residual stress at the surface of the spring wire (Todinov, 2000). During operation, this compressive residual stress offsets applied tensile stress, effectively reducing the overall stress range and significantly increasing fatigue life (Figure 8.7). For example, the fatigue life of leaf springs can be improved tenfold through shot peening.

In the construction industry, pre-tensioning is a prominent example of using an inverse state to buffer detrimental effects. By embedding pre-tensioned tendons (steel cables or rods) in concrete beams, a clamping force is introduced. Once the concrete cures, the beam is preloaded in compression. This compressive residual stress counteracts the tensile stresses that arise from bending during service. As a result, the net stress is either compressive or a significantly reduced tensile stress. Pre-stressed concrete is widely used for constructing floors in high-rise buildings and enables the design of long-span bridges and large column-free spaces in architectural structures.

An inverse state of compressive residual stresses at the surface, acting as a buffer to compensate for tensile service stresses during loading, can be intentionally introduced through specialized heat and thermochemical treatments such as *case-hardening*, *gas carburizing*, and *gas nitriding*.

Other examples of anticipatory inverse states include corrosion, erosion, and wear allowances added to the computed dimensions of components (e.g., pipes, structural

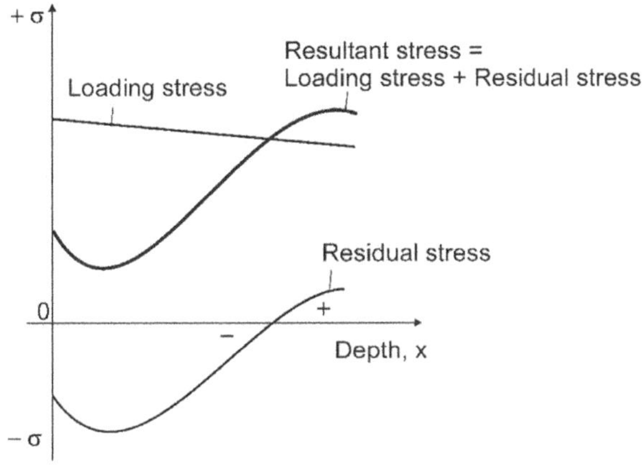

FIGURE 8.7 Resultant stress distribution near the surface of a loaded, shot-peened compression spring.

supports exposed to the elements, cutting tools, etc.). These allowances serve as buffers against wall thickness reduction and significantly lower the risk of failure over time.

The concept of inverse states acting as buffers finds broad application beyond engineering. In project management, time buffers allocated to critical tasks help reduce the risk of delays if unexpected issues arise. Similarly, in inventory management under uncertain demand, increasing the stock of critical items, such as life-saving medications, reduces the risk of stockouts during demand surges and clustering of demands.

In finance, building up financial reserves helps banks and companies remain resilient against credit and market risks. Likewise, in experimental work, preparing additional specimens mitigates the risk of having too few functional samples due to issues such as poor preparation, damage during transport or testing, or defects in the material.

8.4.4 Inverse States Limiting the Likelihood of Erroneous Actions

An example of an inverse state designed to reduce the risk of errors can be found in railway signalling. Rather than activating a signal only in case of danger, the system is configured so that the default state is 'danger.' To allow a train to proceed, this danger signal must be explicitly cleared. If the signal operator becomes incapacitated or the system fails to activate, the train remains stopped by default. This approach ensures that, in the absence of a positive action, the system defaults to a safe state, effectively eliminating dangerous failure modes.

A similar principle applies when users are required to select m items from a total of n options, especially when m is close to n. Instead of starting with an empty selection and adding items until m are selected, the system can begin with all n options pre-selected. The user then deselects the few unwanted items to arrive at the correct set of m. This inverse approach involves fewer steps, which reduces significantly the time for the selection process.

8.5 IMPROVING RELIABILITY AND REDUCING RISK BY INVERSE THINKING

8.5.1 Starting from the Desired End Result

Formulating the desired end result is often a crucial first step in developing reliability-enhancing solutions. By envisioning how a system would behave under ideal conditions, without failures, human error or downtime, engineers can develop more effective and robust solutions.

One such desired end result is a system immune to operator error. To achieve this goal, solutions include automation, error-proofing mechanisms (Poka-Yoke), and interlocks that physically or logically prevent incorrect actions. Additional measures such as permission-based software controls and automatic calibration can further minimize the possibility of human-induced faults.

Another desired end result is the use of 'eternal' components such as parts that experience virtually no wear-out failures. This can be pursued by selecting high-durability materials such as ceramics, composites, or nanomaterials that resist fatigue, corrosion, and thermal stress.

A further desired end result is zero system downtime. This can be achieved by designing systems with hot-swappable components, which allow parts to be replaced or upgraded without interrupting system operation.

In the realm of information systems, perfect data integrity (no corrupted data or communication loss) is a desired end result. This can be achieved through error detection and correction mechanisms, including error-correcting codes, cyclic redundancy checks and secure communication protocols that identify and rectify errors in real time.

Consider the example of a sprinkler system intended to activate when temperatures exceed a critical threshold in the event of fire. A traditional design might involve sensors and an electronic control mechanism to trigger the sprinkler. However, this adds complexity, cost, and introduces a new failure mode: the control system fails precisely when it is needed most.

Instead, a system that activates automatically and reliably at the critical temperature, without electronics or sensors, leads to a simpler, more robust solution. By using a temperature-sensitive alloy with a melting point matching the critical threshold, the system becomes entirely passive. When the alloy melts in the event of fire, it directly triggers the sprinkler, eliminating the need for external control systems. This approach improves reliability, reduces cost, and simplifies the design by minimizing potential failure points.

Another example involves self-healing systems, where the desired end result is a system that automatically detects and corrects its own faults. This can be achieved using self-diagnosing and self-repairing logic, typically through rule-based algorithms or machine learning, which allow the system to identify anomalies, isolate faults, and initiate recovery protocols autonomously, without human intervention.

8.5.2 Inverting the Problem Related to Reliability Improvement and Risk Reduction

Instead of concentrating on how to improve a product's reliability, in this technique, the focus is inverted to explore how the system can be made to fail. Rather than designing for reliability directly, the emphasis shifts to identifying and precipitating failure modes.

The advantage of this inverted approach is that it reframes the problem, offering a fresh perspective that breaks conventional thought patterns and opens new avenues for exploration. Restating the problem from 'how to improve reliability' to 'how to induce failure' often simplifies the problem and uncovers hidden opportunities for improvement.

This problem formulation reversal frequently reveals unexpected resources and strategies for enhancing reliability. Designing for reliability can be abstract and difficult to plan but analysing how failure occurs brings concrete insights and practical solutions. Moreover, when assessing a product, there is often an inherent bias: a tendency to downplay shortcomings and overstate safety. This mindset can obscure subtle, insidious failure modes that pose serious risks to product safety. By deliberately seeking out ways a product might fail, these blind spots are more likely to be uncovered and addressed.

A straightforward example of risk reduction through inverse problem statement is improving protection against unauthorized access to a valuable service controlled by a computer program (e.g., access to a bank account). A powerful approach involves inverting the problem from 'how to defend the system' to 'how to compromise the access-control program and make it fail'.

This reversal of perspective invariably uncovers a range of software vulnerabilities that could be exploited to gain unauthorized access. Common issues revealed through this technique include:

- Intercepting program control after inducing a division-by-zero error,
- Triggering a buffer overflow to hijack execution flow,
- Stealing passwords due to failure to clear the keyboard buffer,
- Inadequate defences against simultaneous, coordinated password attacks.

By focusing on how the access-control program might fail, one uncovers critical weaknesses, making this an essential method for identifying and eliminating software vulnerabilities and reducing risk.

Subversion analysis and anticipatory failure determination, as outlined by Kaplan et al. (1999), are prominent examples of this approach. These techniques emphasize inventing failures using the resources already present in the system. While valuable for identifying rare and unexpected failure modes, they impose a limitation by focusing only on existing resources.

In contrast, the broader method of inverse thinking is not bound by this constraint. It extends beyond available resources, uncovering failure modes that may arise from specific combinations of random events and contextual circumstances. For example, a fatal outcome for a critically ill patient might not stem from a direct fault in lifesaving equipment or personnel, but rather from a clustering of simultaneous demands on those resources. If several critically ill patients require immediate intervention at once, the system may become saturated, leading to a fatal outcome when a new demand emerges.

8.5.3 Improving Reliability and Reducing Risk by Inverting the Focus

For a long time, the inductive approach, which forms the basis of Failure Modes and Effects Analysis (FMEA, MIL-STD-1629A, 1977) has been a fundamental method for identifying potential failure modes at the system level. This technique involves sequentially analysing the failure modes of all parts and components in a system and tracing their effects on overall system performance.

While effective to a degree, this component-level, bottom-up approach has limitations. It cannot uncover all failure modes, particularly those that emerge from system-level interactions or design deficiencies. For example, in a lubrication system, the failure modes 'undersized sump' or 'excessive clearances between lubricated surfaces' cannot be identified solely through a traditional inductive FMEA.

An important enhancement to FMEA involves inverting the focus from components to the system. This system-level perspective begins by examining a known or potential failure mode at the system level and asking: *What lower-level failures could cause or contribute to this?* This inverted focus approach often reveals critical failure modes that the inductive method misses.

For instance, if the system-level failure mode is 'low oil pressure', one can identify contributing causes such as: (i) excessive clearance between lubricated surfaces due to wear, (ii) leaks from seals and gaskets, or (iii) the use of oil with inappropriate (too high) viscosity. Similarly, considering the system-level failure mode 'high oil pressure', might reveal the low-level failure mode 'incorrect design of the oil galleries'.

Likewise, investigating the cause of 'high oil temperature' may uncover the system condition 'small amount of circulating oil' which, in turn, points to low-level failure modes such as 'low oil level' or 'undersized sump'. An undersized sump, by reducing oil volume, leads to insufficient cooling and hence elevated oil temperatures which is a system-level failure mode.

By shifting the analysis from component failures to system-level undesirable effects and backtracking their potential causes, engineers can uncover hidden vulnerabilities and gain a deeper understanding of system reliability. This inverse, top-down thinking complements traditional FMEA and significantly enhances the discovery of failure modes that might otherwise remain undetected.

8.5.4 Enhancing Reliability by Monitoring for the Absence of Expected Events

Improving system reliability and reducing risk often requires a shift in perspective from focusing solely on the occurrence of critical events to recognizing the *absence* of expected signals or activities during operation.

For example, the absence of error logs might initially seem positive. However, if a system never logs anything, it could indicate that the logging mechanism is disabled or malfunctioning. Similarly, missing sensor signals could point to sensor failure, communication issues, or internal system faults. Some components, like filters or bearings, naturally degrade over time. If sensors never report gradual wear, they may be non-functional or providing misleading data.

In semi-automated systems, regular human interactions, such as overrides or approvals, are expected. A lack of these inputs may indicate interface failures, workflow interruptions, or insufficient training.

The absence of scheduled maintenance activities is also a red flag. Components that require periodic alignment, cleaning, lubrication, or testing may be degrading unnoticed. Similarly, the lack of routine part replacements, especially for high-wear items, can increase the risk of unexpected failure.

In manufacturing or procedural workflows, missing step-confirmation signals can allow quality or safety issues to go undetected. Implementing digital checklists or automated step logging can help mitigate these risks.

Finally, engineering systems typically experience varying loads and inputs. If system outputs remain flat or constant, it may suggest that the system is not operating correctly, is receiving no real input, or is logging data inaccurately.

9 Improving the Reliability of Systems by Exploiting Algebraic Inequalities

9.1 REVERSE ENGINEERING OF ALGEBRAIC INEQUALITIES FOR GENERATING NEW KNOWLEDGE AND ENHANCING PERFORMANCE OF SYSTEMS AND PROCESSES

Standard textbooks on reliability engineering and system reliability do not address a fundamental question related to how to improve system reliability without requiring specific knowledge of individual component reliabilities. Even components of the same type exhibit significant uncertainty in reliability due to factors such as manufacturer, age, operating environment, and duty cycle. For example, sensors of the same type sourced from different suppliers can show considerable reliability variations because of the different variety.

Information about the reliability of different component varieties from suppliers is often missing, difficult to obtain, or highly unreliable. Therefore, it is crucial to develop methods for enhancing system reliability without relying on any knowledge of the reliabilities of individual components of different varieties.

Despite the extensive research on redundancy optimization in the reliability literature, no existing technique has yet addressed redundancy optimization without requiring specific knowledge of component reliabilities. Traditionally, optimal redundancy allocation methods begin with known reliability values for all components and redundancies. For example, a multi-performance redundancy optimization technique for multi-state systems using a genetic algorithm has been introduced in (Yi et al. 2021, relying on known component reliabilities. Similarly, Aqel and Mellal (2023) addressed the redundancy allocation problem with nature-inspired AI algorithms, such as adaptive particle swarm optimization, which also depend on known component reliabilities.

To address redundancy allocation in large systems, several approaches have been presented in (Florin et al., 2020), including the Lagrange multipliers technique, a pairwise hill-climbing algorithm, and an evolutionary algorithm. Additionally, Si et al. (2020) reviewed system reliability optimization techniques driven by importance measures, which also depend on known component reliabilities.

The key insight of this chapter is in understanding that by reverse engineering of algebraic inequalities, system reliability can be improved without any knowledge of component reliability values. Specifically, for series-parallel systems, an asymmetric

DOI: 10.1201/9781003679301-9

arrangement of interchangeable redundancies consistently results in higher system reliability compared to a symmetric arrangement, regardless of the individual reliability values of the components or their ranking. This result has been obtained through reverse engineering of a complex algebraic inequality.

For a long time, algebraic inequalities have been employed for estimating upper and lower bounds, quantifying errors, and defining design constraints in engineering applications. Traditionally, inequalities have served as analytical tools. However, a fundamental shift in their application has emerged with the concept of *reverse engineering of algebraic inequalities which is* a novel approach that physically interprets algebraic inequalities to generate new knowledge and optimize engineering systems and processes (Todinov, 2024a, 2024b, 2025a).

The reverse engineering of algebraic inequalities begins with a mathematically valid inequality which is interpreted within a physical or engineering context. When the variables and terms in such an inequality correspond to controlling factors or attributes of a physical system or process, the system exhibits behaviours consistent with the inequality's predictions (Todinov, 2025a). In this way, reverse engineering of algebraic inequalities can reveal new insights and even uncover previously unrecognized physical principles.

Through this process, valid algebraic inequalities become tools for discovering new relationships and enhancing the performance of systems without the need for prior data. The knowledge generated can be applied to optimize design, improve reliability, and guide decision making.

9.2 RELIABILITY OF SYSTEMS WITH COMPONENTS ARRANGED IN SERIES AND PARALLEL

Figure 9.1 depicts a system with components of types C1, C2, ..., Cn, logically arranged in series (Figure 9.1a), and a system with the same components logically arranged in parallel (Figure 9.1b).

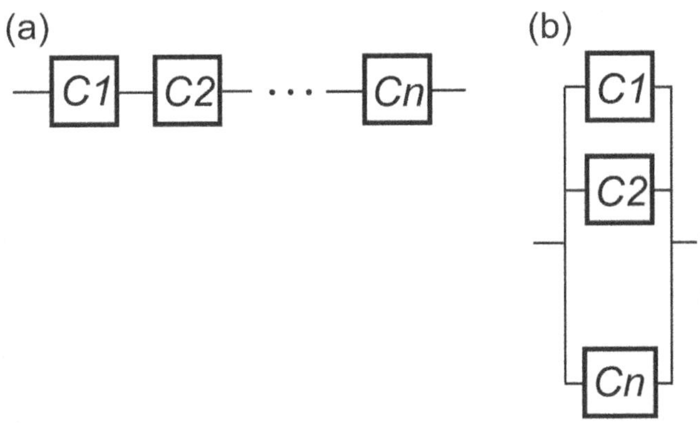

FIGURE 9.1 (a) A system with components logically arranged in series; (b) a system with components logically arranged in parallel.

The system with components in series is in a working state when all components are functioning, while the system with components arranged in parallel is in a working state when at least one of the components is functioning.

Suppose that $r_1, r_2, ..., r_n$ $(0 \leq r_i \leq 1)$ stand for the reliabilities of component types C1, C2, …, Cn. These are the probabilities that the components will be in a working state (functioning). The reliability $R_{a,s}$ of the system in Figure 9.1a is then given by the classical equation (Hoyland and Rausand, 1994; Ebeling, 1997):

$$R_{a,s} = r_1 \times r_2 \times, ..., \times r_n \tag{9.1}$$

In words, the probability that the system will be in a working state is equal to the product of the probabilities that the components building the system will be in a working state.

In contrast, $f_1 = 1 - r_1$; $f_2 = 1 - r_2; ...; f_n = 1 - r_n$ $(0 \leq f_i \leq 1)$ stand for the probabilities that the components will be in a failed state.

The probability $F_{a,p}$ that the system with components logically arranged in parallel (Figure 9.4b) will be in a failed state is given by (Ebeling, 1997):

$$F_{a,p} = f_1 \times f_2 \times, ..., \times f_n \tag{9.2}$$

which can also be presented as:

$$F_{a,p} = (1 - r_1) \times (1 - r_2) \times ... \times (1 - r_n) \tag{9.3}$$

In words, the system with components arranged in parallel is in a failed state when all components building the system are in a failed state.

The reliability $R_{a,p}$ of the system with components arranged in parallel is given by $R_{a,p} = 1 - F_{a,p}$ or

$$R_{a,p} = 1 - (1 - r_1) \times (1 - r_2) \times ... \times (1 - r_n) \tag{9.4}$$

9.3 A GENERAL ALGEBRAIC INEQUALITY

Suppose that $0 \leq r_1, r_2, ... r_m \leq 1$ are non-negative numbers with values between 0 and 1. Then, it can be shown that the general inequality

$$(1 - r_1^n)(1 - r_2^n)...(1 - r_m^n) \leq (1 - c_{11}c_{21}...c_{n1})(1 - c_{12}c_{22}...c_{n2})...(1 - c_{1m}c_{2m}...c_{nm}) \tag{9.5}$$

holds. Any of the $c_{ij} \geq 0$ in the right-hand side of inequality (9.5) is equal to one of the values $r_1, r_2, ..., r_m$, characterizing the reliabilities of the m different types of components. In addition, in the right-hand side of 9.5 there are exactly n numbers c_{ij} equal to r_1, exactly n numbers c_{ij} equal to r_2, …, and exactly n numbers c_{ij} equal to r_m. The proof of the inequality (9.5) has been given in (Todinov, 2025b).

9.4 IMPROVING THE RELIABILITY OF PARALLEL SYSTEMS THROUGH SYMMETRIC ARRANGEMENT OF COMPONENTS

In what follows, applications related to improving the reliability of common parallel systems with interchangeable components of the same type but of different varieties will be proposed. The reliability improvement method is low-cost and domain-independent and can be used for improving reliability and reducing the risk of failure in various unrelated domains. This method is derived from the reverse engineering of algebraic inequality (9.5).

The algebraic inequality:

$$\begin{aligned}&(1-r_1^2)(1-r_2^2)...(1-r_m^2) \le (1-r_1 r_2)(1-r_2 r_3)...(1-r_m r_1)\\&0 \le r_i \le 1\end{aligned} \tag{9.6}$$

where r_i are variables accepting values from the interval [0, 1], is a special case of inequality (9.5), for n=2. As can be verified, in the right-hand side of inequality 9.6, there are exactly n=2 numbers equal to r_1, exactly n=2 numbers equal to r_2, …, and exactly n=2 numbers equal to r_m.

Now, let r_i, $i = 1,...,m$ in (9.6) be physically interpreted as the 'reliabilities' of components working independently from one another, from varieties R_i $(i = 1,...,m)$ (Figure 9.2). Let us suppose that all components are of the same type (e.g., type 'seals') but of different variety (e.g., from a different manufacturer).

The left-hand side of inequality (9.6) can be physically interpreted as the probability of system failure $F_b = (1-r_1^2)(1-r_2^2)(1-r_3^2)...(1-r_m^2)$ for the configuration

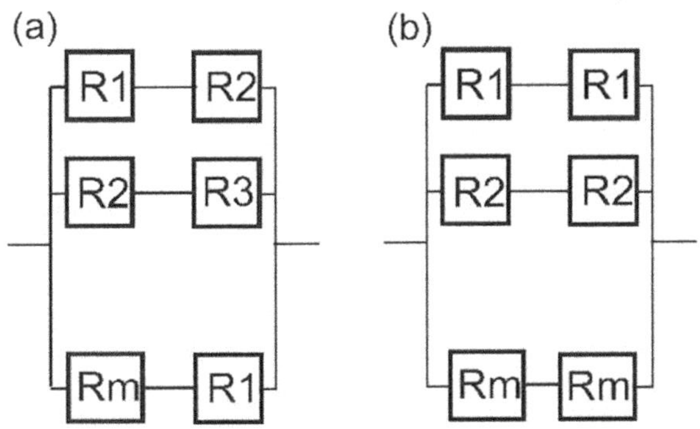

FIGURE 9.2 A reliability network of a parallel system with interchangeable components of different varieties with (a) asymmetric and (b) symmetric arrangements of the interchangeable components.

shown in Figure 9.2b, while its right-hand side corresponds to the probability of failure $F_a = (1-r_1r_2)(1-r_2r_3)(1-r_3r_4)...(1-r_mr_1)$ for the system depicted in Figure 9.2a. Indeed, the systems in Figure 9.2 are in a working state if at least one of the parallel branches is in a working state. The systems are in a failed state when all parallel branches are in a failed state. Therefore, the probability of failure of the parallel systems in Figure 9.2 is equal to the product of the probabilities of failure of every single parallel branch in the systems, which yields the expressions $F_a = (1-r_1r_2)(1-r_2r_3)(1-r_3r_4)...(1-r_mr_1)$ and $F_b = (1-r_1^2)(1-r_2^2)(1-r_3^2)...(1-r_m^2)$ for the probabilities of failure of the systems in Figure 9.2a and Figure 9.2b, correspondingly.

The reverse engineering of inequality (9.6) effectively establishes that the symmetric arrangement of interchangeable components of the same variety in each parallel branch (Figure 9.2b) is always characterized by a smaller probability of system failure, compared with their asymmetric arrangement (Figure 9.2a), *irrespective of the reliabilities* r_i *of the components.*

Note that even though the reliabilities r_i of component varieties and their rankings are unknown, the reliability of the system in Figure 9.2a, which has an asymmetrical arrangement of the component varieties, can still be enhanced by introducing a symmetric arrangement of the component varieties (Figure 9.2b). *To improve the system's reliability, there is no need to know the reliabilities of the different varieties or their ranking according to their reliability.*

We must point out that no ranking of the component varieties by their reliability is necessary for the method to be applied. To demonstrate this, we select the parallel systems in Figure 9.3 consisting of four branches connected in parallel.

The reliabilities of the component varieties and the probabilities of failure of the systems in (a) and (b) are given in Table 9.1. The values r_1, r_2, r_3, and r_4 correspond to the reliabilities of the component varieties R_1, R_2, R_3, and R_4 in Figure 9.3. The values $F_a = (1-r_1r_2)(1-r_2r_3)(1-r_3r_4)(1-r_4r_1)$ are the probabilities of failure of the systems with an asymmetric arrangement of the component varieties (Figure 9.3a), while the values $F_b = (1-r_1^2)(1-r_2^2)(1-r_3^2)(1-r_4^2)$ are the probabilities of failure of the system with a symmetric arrangement of the interchangeable component varieties (Figure 9.3b).

As evidenced by Table 9.1, in all permutations of component reliabilities, the symmetric arrangement of component varieties across the parallel branches consistently results in a lower probability of system failure ($F_b < F_a$). This trend holds universally, with no exceptions observed, even if the table were extended indefinitely.

To maximize the reliability of the original system, under conditions of complete uncertainty regarding the reliabilities of individual components, the system's configuration must eliminate all asymmetries throughout its structure.

Let us consider the system shown in Figure 9.4a. Its reliability can be improved by making the arrangement of component varieties symmetrical, in the first and second branches. This arrangement results in the modified system depicted in Figure 9.4b.

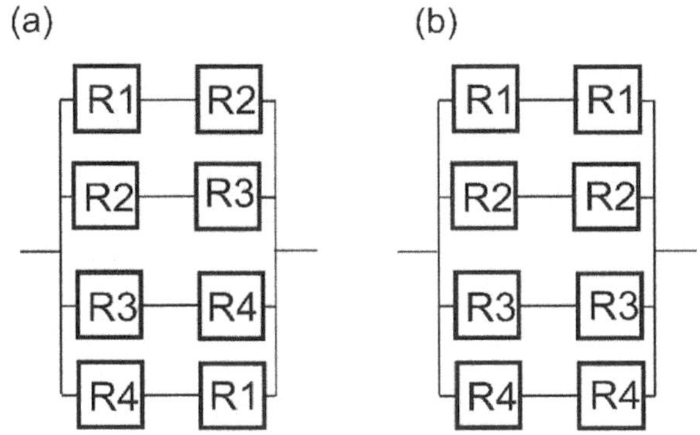

FIGURE 9.3 (a) A system with an asymmetric arrangement of the interchangeable components; (b) a system with a symmetric arrangement of the interchangeable components.

TABLE 9.1
The reliabilities of the systems in Figure 9.2 for different values of the reliabilities of the component varieties forming the systems

r_1	r_2	r_3	r_4	F_a	F_b
0.61	0.92	0.75	0.45	0.065	0.034
0.2	0.6	0.1	0.7	0.662	0.31
0.8	0.4	0.65	0.22	0.355	0.166
0.1	0.75	0.65	0.12	0.432	0.246
0.85	0.55	0.88	0.46	0.0996	0.0344
0.5	0.7	0.25	0.7	0.288	0.183
0.35	0.48	0.70	0.14	0.474	0.338
0.1	0.2	0.15	0.7	0.791	0.474
0.89	0.99	0.79	0.95	9.987×10^{-4}	1.51×10^{-4}

However, further improvements can be achieved by eliminating the remaining asymmetries in the third and fourth parallel branches. This leads to the fully symmetric configuration illustrated in Figure 9.4c. The system in Figure 9.4c is more reliable than the one in Figure 9.4b for the following reasons:

In Figure 9.4b, the section shown in Figure 9.4d is connected in parallel within the system, whereas in Figure 9.4c, the corresponding section is that shown in Figure 9.4e.

The section in Figure 9.4d exhibits a higher probability of failure compared with the symmetrical and more reliable section in Figure 9.4e.

Interchangeable components may encompass various types that align with real-world applications. For instance, interchangeable temperature gauges can operate based on different physical principles such as thermocouples, radiation thermometers, thermal resistors, or bi-metal thermometers. Similarly, a wide range of options exist

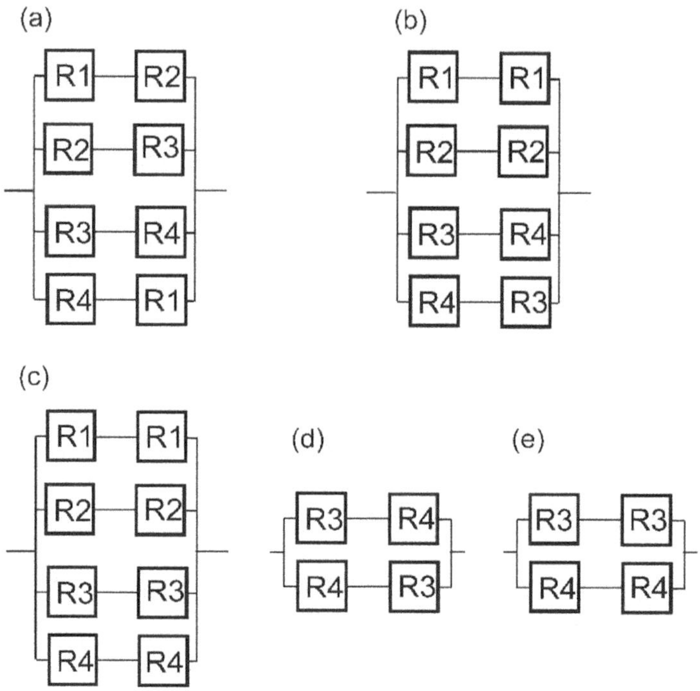

FIGURE 9.4 (a) Initial system with components arranged in parallel; (b) a system with improved reliability obtained from the initial system by destroying the asymmetry in part of the system; (c) a system with maximized reliability, obtained by destroying the asymmetry in all parts of the initial system (a); (d) a subsystem, connected in parallel within the system (b); (e) a subsystem, connected in parallel within the system (c).

for measuring other parameters, like pressure, concentration, and more. Moreover, even within a single component type, variations in component reliability can arise depending on the manufacturer and the age of the components.

It needs to be clarified that the interchangeable component varieties and types can be characterized with distinct failure modes, and these do not restrict the application of the proposed method. The distinct failure modes (e.g., wear-out failures and overstress failures) affect only the time to failure of the components and their reliability values.

9.5 APPLICATION EXAMPLES RELATED TO IMPROVING RELIABILITY OF PARALLEL SYSTEMS

An example of a mechanical system that can benefit from a symmetric rearrangement of components of the same variety in the parallel branches is given in Figure 9.5. The system consists of three parallel lines transporting cooling fluid from sources s1, s2, and s3 to a chemical reactor CR with two pumping sections on each parallel line.

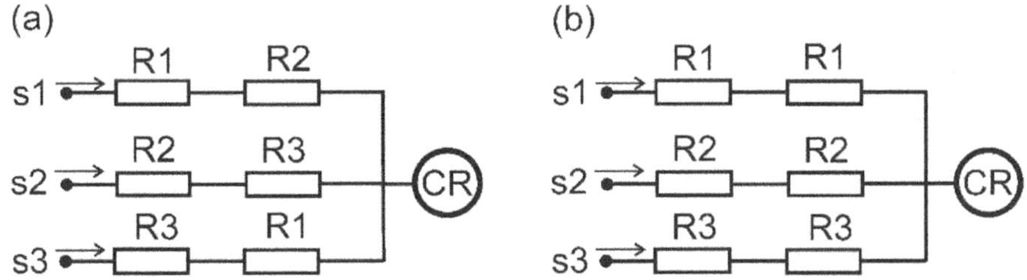

FIGURE 9.5 A system of three pipelines transporting cooling fluid to a chemical reactor CR with two pumping stations on each branch: (a) an asymmetric arrangement of the pumping station types in the branches; (b) a symmetric arrangement of the pumping station types in the branches.

The system is deemed operational, if at least one of the parallel branches is operational and delivers cooling fluid to the chemical reactor. The interchangeable pumping sections are of varieties R1 (a pumping section from manufacturer '1', R2 (a pumping station from manufacturer '2'), and R3 (a pumping station from manufacturer '3'), characterized by unknown reliabilities r_1, r_2, r_3, respectively.

The reliability networks of the systems in Figure 9.5a,b are given in Figure 9.2a,b respectively, for $m = 3$ parallel branches. These parallel systems are quite common in numerous engineering applications.

The proposed reliability improvement technique exploiting the reverse engineering of inequality (9.6) is domain independent. It can be used for any parallel system with interchangeable components: valves, sensors, seals, etc.

In another example, Figure 9.6 presents a system that can also benefit from a symmetric arrangement of interchangeable sensors in the parallel branches.

This system comprises two zones, *A* and *B*, where the temperature (pressure, concentration, etc.) is measured in each zone by interchangeable sensors of varieties *X* and *Y*. To enhance the system's reliability, two parallel measurement channels, Ch1 and Ch2, have been introduced, each measuring the difference in temperature/pressure/concentration between the two zones (Figure 9.6a). To guarantee operational safety, at least one of the channels must provide a valid measurement from each zone. Channel redundancy ensures that a valid measurement of the temperature difference between the two zones will still be made even if any of the sensors fails.

The sensors are of varieties *X* and *Y*, characterized by unknown reliabilities r_1 and r_2, respectively. The reliability networks of the systems in Figure 9.6a,b are given in Figures 9.6c,d, respectively. These are also parallel systems with interchangeable components. For $m = 2$, inequality (9.6) transforms into the special-case inequality

$$\begin{aligned} &(1 - r_1^2)(1 - r_2^2) \le (1 - r_1 r_2)(1 - r_2 r_1) \\ &0 \le r_i \le 1 \end{aligned} \tag{9.7}$$

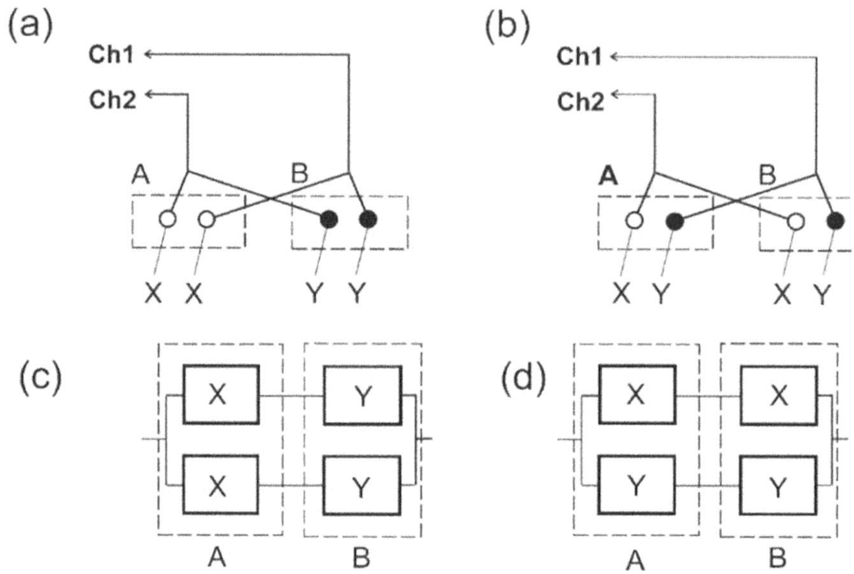

FIGURE 9.6 A system of interchangeable sensors of varieties *X* and *Y*, measuring the temperature/pressure/concentration in two zones, *A* and *B*: (a) an asymmetric arrangement of the sensors in each channel; (b) a symmetric arrangement of the sensors in each channel; (c) the reliability network of the system in (a); (d) the reliability network of the system in (b).

The left-hand side, $F_b = (1 - r_1^2)(1 - r_2^2)$, of inequality (9.7) can be physically interpreted as the probability of failure F_b of the system in Figure 9.6b, while the right-hand side, $F_a = (1 - r_1 r_2)(1 - r_2 r_1)$, of inequality (9.7) can be physically interpreted as the probability of failure F_a of the system in Figure 9.6a.

According to inequality (9.7), the probability of failure of the system with the same variety of sensors in each channel (Figure 9.6b) is smaller than the probability of failure of the system with different varieties of sensors in each channel (Figure 9.6a) ($F_b \le F_a$).

The difference in the probabilities of failure of the systems in Figures 9.6a,b can be significant, as the next numerical examples demonstrate.

Let us suppose that for a given period of operation, the reliabilities r_1 and r_2 of the sensor varieties are those in Table 9.2. The table also lists the probabilities of failure $F_a = (1 - r_1 r_2)(1 - r_2 r_1)$ and $F_b = (1 - r_1^2)(1 - r_2^2)$ of the systems in Figures 9.6a,b. The probabilities of failure of the systems in Figure 9.6a are significantly greater than the probabilities of failure of the systems in Figure 9.6b. The symmetric configuration of the sensor varieties in Figure 9.6b leads to a dramatic improvement in the system's reliability. Table 9.2 summarizes these results.

TABLE 9.2
The reliabilities of the systems in Figure 9.6 for different values of the reliabilities of the sensors

r_1	r_2	F_a	F_b
0.16	0.85	0.75	0.27
0.76	0.4	0.48	0.35
0.9	0.2	0.67	0.18
0.85	0.32	0.53	0.24
0.09	0.7	0.88	0.5
0.1	0.7	0.865	0.505
0.98	0.94	6.21×10^{-3}	4.61×10^{-3}

9.6 PARALLEL SYSTEMS WITH MULTIPLE COMPONENTS IN EACH BRANCH

The reverse engineering of algebraic inequalities can also enhance the reliability of parallel systems composed of multiple components per branch, particularly when the reliabilities of individual components are unknown. It is assumed that each parallel branch in the system contains the same number of interchangeable components.

Let us consider the correct abstract algebraic inequality

$$(1-x^3)(1-y^3)(1-z^3) \le (1-x^2y)(1-y^2z)(1-z^2x) \tag{9.8}$$

which is a special case of the general inequality (9.5), for m=3 and n=3. In inequality (9.8): $0 \le x \le 1$, $0 \le y \le 1$, and $0 \le z \le 1$.

Let the variables x, y, and z in inequality (9.8) represent the reliabilities of components of types X, Y, and Z, respectively. In this context, the left-hand and right-hand sides of inequality (9.8) can be physically interpreted as the probabilities of failure of two alternative parallel systems. To illustrate this interpretation, let us consider the physical configurations depicted in Figure 9.7. Each system comprises three parallel gas lines, where each line consists of pumping sections of types X, Y, and Z, connected in series. The entire system operates successfully if at least one gas line can deliver gas to the consumer, C.

The gas lines in the system in Figure 9.7a have an asymmetric arrangement of the types of pumping stations, while the gas lines in the system in Figure 9.7b have a symmetric arrangement of the types of pumping stations.

In relation to gas delivery to consumer C, the reliability networks associated with the physical setups in Figure 9.7a,b are shown in Figure 9.8a and Figure 9.8b, respectively. These networks illustrate the logical arrangement of the systems' components.

If the reliabilities of the section types X, Y, and Z are represented by x, y, and z, respectively ($0 \le x, y, z \le 1$), then the probability of failure F_b of the system in Figure 9.7b is given by $F_b = (1-x^3)(1-y^3)(1-z^3)$, whereas the probability of

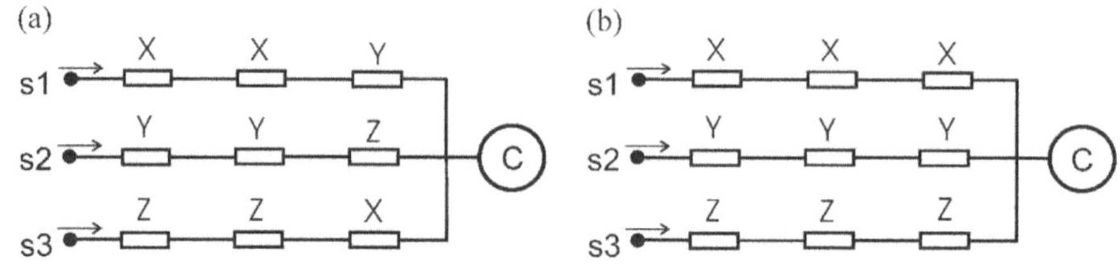

FIGURE 9.7 Functional diagrams of two different arrangements of the interchangeable pumping sections on three pipelines: (a) an asymmetric arrangement of the pumping sections; (b) a symmetric arrangement of the pumping sections.

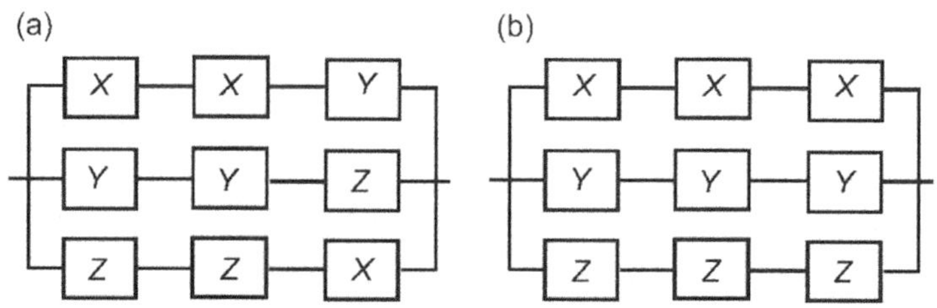

FIGURE 9.8 Reliability networks of the systems in Figure 9.7: (a) reliability network of the system in Figure 9.7a; (b) reliability network of the system in Figure 9.7b.

failure F_a for the system in Figure 9.7a is given by $F_a = (1 - x^2 y)(1 - y^2 z)(1 - z^2 x)$. The reliabilities x, y, and z, of the pumping sections and their rankings are unknown.

By reverse engineering of inequality (9.8), it can be inferred that its left-hand side corresponds to the probability of failure for the system shown in Figure 9.7b, while the right-hand side reflects the probability of failure for the system in Figure 9.7a.

As stated in inequality (9.8), the system depicted in Figure 9.7b exhibits higher reliability than the one in Figure 9.7a. This conclusion is drawn *without any prior knowledge of the failure probabilities x, y, and z for the three distinct section types or of their relative ranking.*

Inequalities similar to inequality (9.5) can be reverse engineered relatively easily because products of the type $(1 - r_1^n)(1 - r_2^n)...(1 - r_m^n)$, where r_i is the reliability of component of type i, can be physically interpreted directly as the probability of failure of a parallel system with m parallel branches and n components of the same variety in each branch.

In this connection, inequalities (9.9)–(9.11), where $0 \le x, y, z \le 1$, are also special cases of the general inequality (9.5). Inequalities (9.9)–(9.11) can also be reverse engineered as 'probabilities of failure of parallel systems with multiple interchangeable components'.

$$(1 - x^3)(1 - y^3)(1 - z^3) \le (1 - xyz)(1 - yzx)(1 - zxy) \tag{9.9}$$

$$(1-x^4)(1-y^4)(1-z^4) \le (1-x^2y^2)(1-y^2z^2)(1-z^2x^2) \tag{9.10}$$

$$(1-x^4)(1-y^4)(1-z^4) \le (1-x^2yz)(1-y^2zx)(1-z^2xy) \tag{9.11}$$

By physically interpreting x, y, and z as the reliabilities of components from varieties X, Y, and Z, respectively, the reverse engineering of inequalities (9.9)–(9.11) reveals that their left- and right-hand sides represent the probabilities of failure of the parallel systems depicted in Figures 9.9–9.11. Specifically, the left-hand sides of the inequalities correspond to the reliability of the systems in Figure 9.9b, Figure 9.10b, and Figure 9.11b, respectively, while the right-hand sides correspond to the reliability of the systems in Figure 9.9a, Figure 9.10a, and Figure 9.11a, respectively.

The reliability improvement technique obtained for parallel systems with an equal number of components in each parallel branch cannot be automatically extrapolated to systems with a different number of components in each parallel branch because systems with a different number of components in each branch are no longer symmetrical. If, for systems with an unequal number of components in each branch, inequalities similar to inequalities (9.8)–(9.11) can be proved to be correct, then symmetric arrangements of components will result in a significant system reliability improvement.

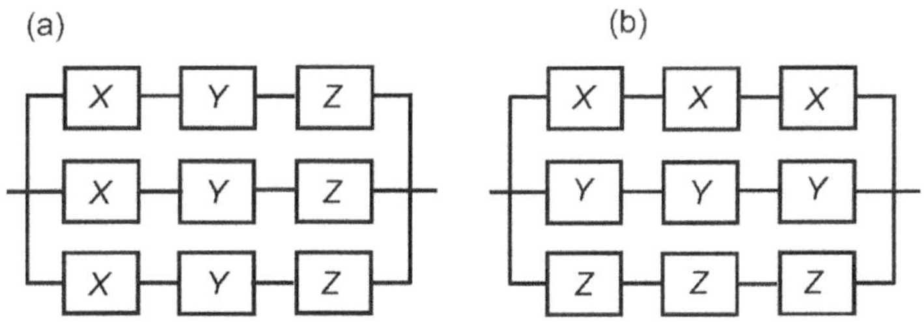

FIGURE 9.9 (a) A reliability network obtained from the physical interpretation of the right-hand side of inequality (9.9) and (b) a reliability network obtained from the physical interpretation of the left-hand side of inequality (9.9).

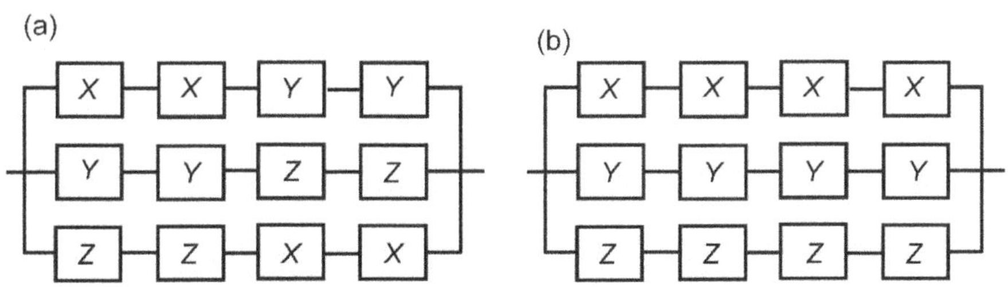

FIGURE 9.10 (a) A reliability network obtained from the physical interpretation of the right-hand side of inequality (9.10) and (b) a reliability network obtained from the physical interpretation of the left-hand side of inequality (9.10).

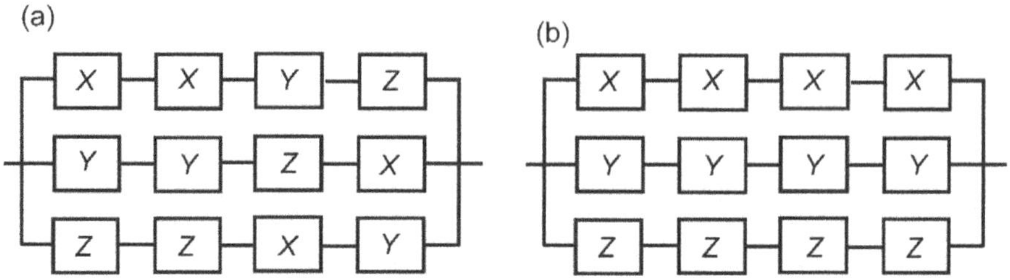

FIGURE 9.11 (a) A reliability network obtained from the physical interpretation of the right-hand side of inequality (9.11) and (b) a reliability network obtained from the physical interpretation of the left-hand side of inequality (9.11).

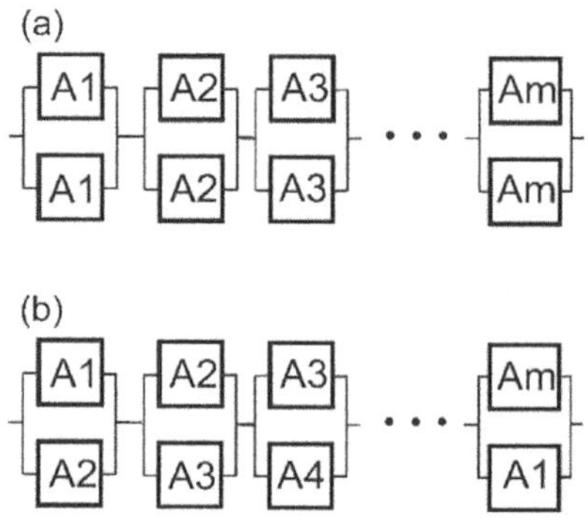

FIGURE 9.12 Reliability network with interchangeable redundancies with (a) a symmetric and (b) an asymmetric arrangement of the interchangeable redundancies.

9.7 IMPROVING THE RELIABILITY OF SERIES-PARALLEL SYSTEMS THROUGH ASYMMETRIC ARRANGEMENT OF COMPONENTS

Interestingly, inequality (9.5) can also be reverse engineered to represent series–parallel systems (Figure 9.12a) if the variables in the inequality are interpreted not as reliabilities but as probabilities of failure.

Let us rewrite inequality (9.6), which is a special case of inequality (9.5), in the following form:

$$\begin{aligned}&(1-f_1^2)(1-f_2^2)...(1-f_m^2) \le (1-f_1 f_2)(1-f_2 f_3)...(1-f_m f_1)\\&0 \le f_i \le 1\end{aligned} \tag{9.12}$$

where f_i are variables accepting values from the interval [0, 1].

Now, let f_i, $i = 1,...,m$ in (9.12) be physically interpreted as 'probabilities of failure' of components of type A_i $(i = 1,...,m)$. The reverse engineering of inequality (9.12) then yields that the left-hand side of the inequality can be physically interpreted as the reliability $R_a = (1-f_1^2)(1-f_2^2)(1-f_3^2)...(1-f_m^2)$ of the system in Figure 9.12a, while the right-hand side of the inequality $R_b = (1-f_1f_2)(1-f_2f_3)(1-f_3f_4)...(1-f_mf_1)$ can be physically interpreted as the reliability of the system in Figure 9.12b.

Inequality (9.12) effectively establishes the opposite of what has been established for parallel systems: *for series-parallel systems, the asymmetric arrangement of the interchangeable redundancies is characterized by a larger reliability, compared with their symmetric arrangement, irrespective of the probabilities of failure* f_i *of the components*.

Note that even though the probabilities of failure f_i of components and their rankings are unknown, the reliability of the system in Figure 9.12a, which has a symmetrical arrangement of the redundant components, can still be enhanced by introducing *an asymmetric arrangement of redundancies. To improve the system's reliability, there is no need to know the probabilities of failure or the ranking of the components by their probability of failure.*

To demonstrate this, we select the series–parallel system in Figure 9.12 consisting of four sections connected in series (*m*=4).

The probabilities of failure of the components and the reliabilities of the systems in configurations (a) and (b), are given in Table 9.3. The values f_1, f_2, f_3, and f_4 correspond to the probabilities of failure of components A_1, A_2, A_3 and A_4 in Figure 9.12 (*m*=4). The values $R_a = (1-f_1^2)(1-f_2^2)(1-f_3^2)(1-f_4^2)$ are the reliabilities of the systems with symmetric redundancies, while the values $R_b = (1-f_1f_2)(1-f_2f_3)(1-f_3f_4)(1-f_4f_1)$ are the reliabilities of the systems with asymmetric redundancies.

As evidenced by Table 9.3, in every instance of ordering the component probabilities of failure, the asymmetric arrangement of redundancies consistently results in a system with superior reliability ($R_b > R_a$). This trend holds without exception and will continue to do so even if the table is extended indefinitely.

To maximize the reliability of the original system, in the presence of a total uncertainty about the probabilities of failure of the individual components, the arrangement of redundancies must be such that the original symmetric arrangement is destroyed in all parts of the system.

Consider the system in Figure 9.13a. The reliability of this system can be increased by swapping the redundancies in the first and second sections. This results in the system shown in Figure 9.13b. However, the reliability of the system in Figure 9.13b can be further improved by destroying the existing symmetry in the third and fourth sections connected in series. This leads to the system shown in Figure 9.13c.

Indeed, the system in Figure 9.13c has greater reliability than the system in Figure 9.13b because: (i) the section in Figure 9.13d is connected in series in the system from Figure 9.13b, and the section in Figure 9.13e is connected in series in the system from Figure 9.13c; (ii) the section in Figure 9.13d has a smaller reliability than the section in Figure 9.13e.

Interchangeable redundancies can involve different types of components, which correspond to real-world applications. For example, interchangeable redundancies of different types might include interchangeable seals of various kinds, such as O-ring seals, cartridge seals, pressurized and unpressurized seals, etc. Interchangeable

TABLE 9.3
The reliabilities of the systems in Figure 9.12 (m = 4) for different values of the probabilities of failure of the components in the systems

f_1	f_2	f_3	f_4	R_a	R_b
0.7	0.3	0.87	0.43	0.092	0.255
0.32	0.56	0.45	0.68	0.264	0.333
0.88	0.20	0.40	0.91	0.031	0.131
0.21	0.35	0.12	0.54	0.585	0.830
0.56	0.23	0.78	0.39	0.216	0.389
0.75	0.42	0.45	0.88	0.0648	0.114
0.1	0.7	0.25	0.19	0.456	0.716
0.98	0.89	0.82	0.91	4.63×10^{-4}	9.48×10^{-4}

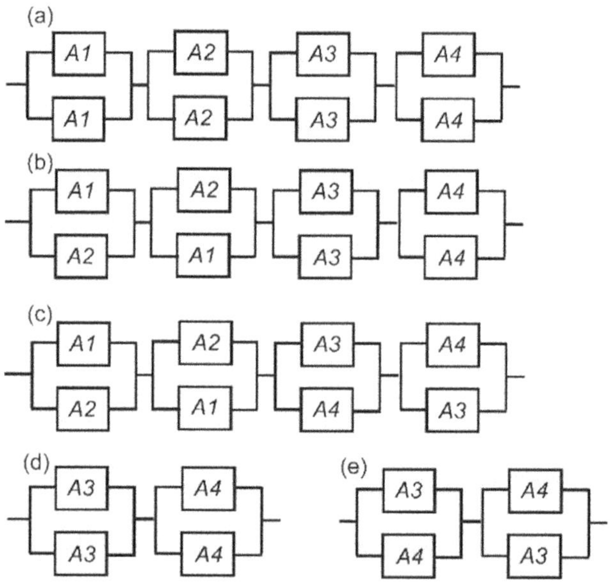

FIGURE 9.13 a) Initial series-parallel system; (b) a system with improved reliability; (c) a system with maximized reliability obtained by destroying the symmetry in all parts of the initial system; (d) a subsystem, connected in series within system (b); (e) a subsystem, connected in series within system (c).

sensors could measure the same quantity but operate on different physical principles. For instance, the temperature in two zones could be measured with thermocouples of different types, thermal resistors, or bi-metal thermometers. Similarly, the pressure in two zones could be measured by bellows-type pressure gauges, manometer pressure gauges based on liquid columns, piezometer pressure gauges, Bourdon tube pressure gauges, and capsule pressure gauges.

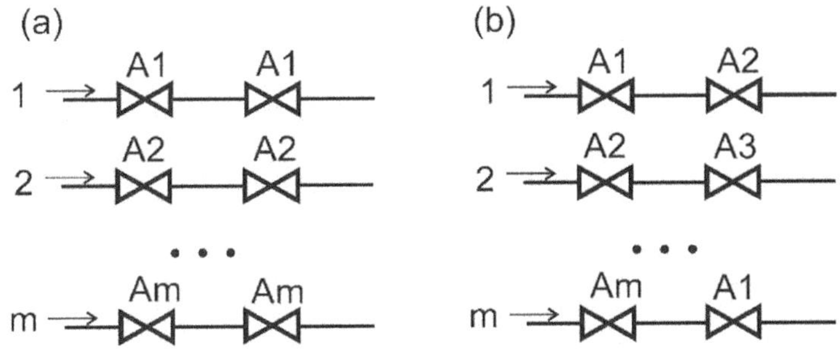

FIGURE 9.14 A system of m pipelines transporting toxic fluid with two valves of types A1, A2, ..., Am, on each pipeline: (a) symmetric arrangement of the redundant valves and (b) asymmetric arrangement of the redundant valves.

Interchangeable redundancies can also involve components of the same type but of different varieties if they are produced by different manufacturers. Components of the same type but different varieties differ in their reliability values. Typically, new components of a particular type have greater reliability than older components of the same type. Similarly, components of the same type produced by different manufacturers also differ in their reliability.

An example of a mechanical system that can benefit from asymmetric arrangement of the redundancies is given in Figure 9.14. The system consists of n pipelines transporting toxic fluid with two valves on each pipeline. All valves are initially open and a signal to close is sent to all of them to stop the fluid in each pipeline.

The system is deemed operational when, upon receiving a command for closure, the flow is halted in all m pipelines. To enhance system reliability, each pipeline features a redundant valve. This redundancy implies that at least one valve on each pipeline must respond to the closure command to ensure that the flow in the pipeline is halted.

The valves are of types A1, A2, ..., Am, characterized by probabilities of failure to close on demand $f_1, f_2, ..., f_m$, correspondingly.

The reliability network of the system in Figure 9.14a is given in Figure 9.12a while the reliability network of the more reliable system in Figure 9.14b is given in Figure 9.12b. These are series-parallel systems and are quite common in numerous engineering applications.

The proposed reliability-improving technique exploiting asymmetry is domain independent. It can be used for any system with interchangeable redundant components: valves, sensors, seals, etc.

9.8 CASE STUDY: IMPROVING THE RELIABILITY OF SERIES-PARALLEL SYSTEMS BY ASYMMETRIC ARRANGEMENT OF REDUNDANCIES

Figure 9.15a presents a mechanical system that can benefit from asymmetric arrangement of redundancies.

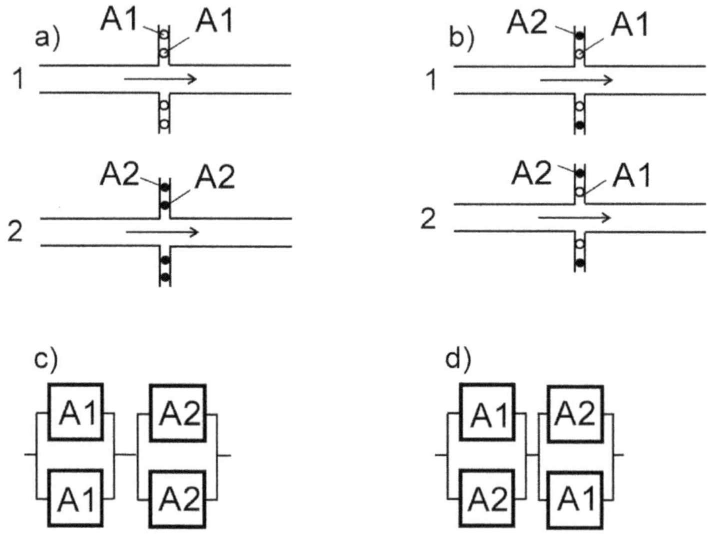

FIGURE 9.15 A system of two pipelines with seals of types A1 and A2: (a) symmetric arrangement of the redundant seals, (b) asymmetric arrangement of the redundant seals, (c) reliability network of system (a); (d) reliability network of system (b).

This system includes two pipelines that transport toxic fluid. Each pipeline is equipped with a flange and each flange has two seals. For the system to be operational, at least one seal on each flange must prevent the toxic fluid from leaking out. To enhance the system's reliability, one of the seals on each flange is redundant. This redundancy ensures that the flow within the pipeline remains isolated, provided at least one seal on each flange is functional.

The seals in the flanges are of types A1 and A2, characterized by probabilities of failure on demand f_1 and f_2, correspondingly. The reliability networks of the systems in Figure 9.15a and 9.15b are given in Figure 9.15c and 9.15d, correspondingly. These are also series-parallel systems with interchangeable redundancies. For $m = 2$, inequality (9.12) gives the special-case inequality

$$\begin{aligned}&(1-f_1^2)(1-f_2^2) \le (1-f_1 f_2)(1-f_2 f_1)\\&0 \le f_i \le 1\end{aligned} \quad (9.13)$$

The left-hand side $R_a = (1-f_1^2)(1-f_2^2)$ of inequality (9.13) can be physically interpreted as the reliability R_a of the seals assembly in Figure 9.15a while the right-hand side $R_b = (1-f_1 f_2)(1-f_2 f_1)$ of inequality (9.13) can be physically interpreted as the reliability R_b of the seals assembly in Figure 9.15b.

According to inequality (9.13), the reliability of the system with mixed types of seals in Figure 9.15b is superior to the reliability of the system with the same type of seals in Figure 9.15a ($R_b \ge R_a$).

The difference in the reliability of the systems in Figure 9.15a and 9.15b can be significant as the next numerical example shows.

Suppose that, for a period of operation of 5 years, the probabilities of failure characterizing the seals are $f_1 = 0.16$ and $f_2 = 0.85$, correspondingly. The system reliabilities are then:

$$R_a = (1 - f_1^2)(1 - f_2^2) = (1 - 0.16^2)(1 - 0.85^2) = 0.27$$

$$R_b = (1 - f_1 f_2)(1 - f_2 f_1) = (1 - 0.16 \times 0.85)(1 - 0.85 \times 0.16) = 0.75$$

The reliability of the seals arrangement in Figure 9.15b is 2.78 times greater than the reliability of the seals arrangement in Figure 9.15a. The asymmetric configuration of seals in Figure 9.15b leads to a dramatic improvement in the system's reliability.

Similar results are obtained for different combinations of values for f_1 and f_2. Table 9.4 summarizes these results.

TABLE 9.4
Reliabilities of the systems in Figure 9.15 for different values of the probabilities of failure of the seals

f_1	f_2	R_a	R_b
0.16	0.85	0.27	0.75
0.76	0.4	0.35	0.48
0.9	0.2	0.18	0.67
0.85	0.32	0.25	0.53
0.09	0.7	0.5	0.88

10 Improving Reliability and Reducing Risk by Exploiting Asymmetry

10.1 CLASSIFICATION OF TECHNIQUES FOR IMPROVING RELIABILITY AND REDUCING RISK BY EXPLOITING ASYMMETRY

The idea of the generic method for reducing risk by exploiting asymmetry is *to reduce the risk associated with a failure mode by leveraging asymmetric positioning, arrangement, geometry, output, cost of failure, or payoffs.*

In standard design literature, there is a notable absence of domain-independent methods for improving the reliability of designed components by exploiting asymmetry. Although asymmetry has already been utilized to enhance the functionality of designs and processes (Kolitsidas et al., 2014; Ilic et al., 2016; Gu et al., 2018; Wang et al., 2018; Coria et al., 2020), the existing approaches exploiting asymmetry are domain specific. Domain-specific methods, however, cannot transcend the narrow domain they serve and cannot normally be used to improve reliability in other, unrelated domains.

Improving the reliability of designs through exploiting asymmetry involves leveraging the asymmetric response from altering the arrangement of features, position, location, orientation, output characteristics, geometry, and properties to achieve better reliability. This often leads to simple, low-cost solutions. As demonstrated in Chapter 9, the domain-independent asymmetric arrangement of redundancies in series-parallel systems can significantly enhance their reliability even in the complete absence of information about the reliability of individual components or their ranking.

In reliability engineering texts, there is a lack of classification of generic, domain-independent techniques for improving reliability by exploiting asymmetry, and the associated reliability-improving mechanisms. Such a classification and discussion of mechanisms have only been recently presented in (Todinov, 2025d). The classification, presented in Figure 10.1, promotes disciplined thinking and prevents the oversight of effective reliability-enhancing solutions exploiting asymmetry.

DOI: 10.1201/9781003679301-10

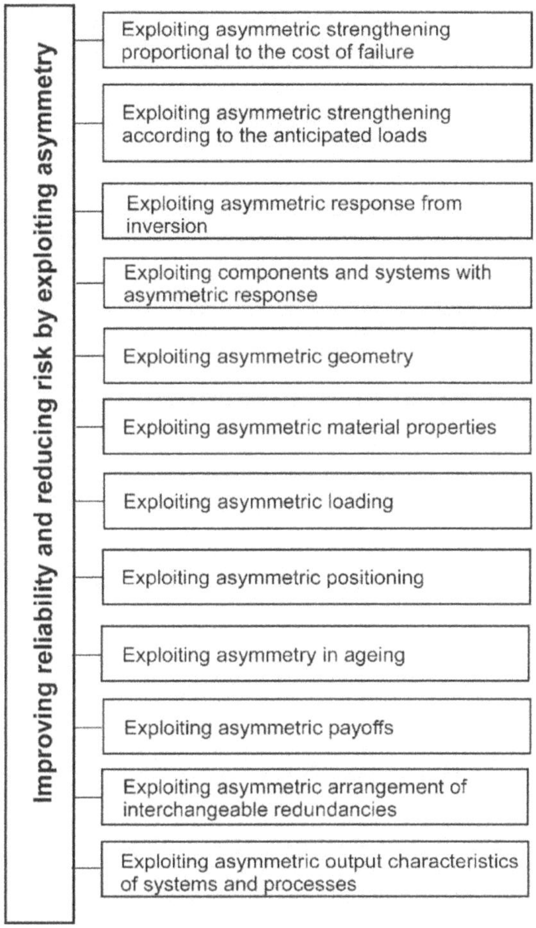

FIGURE 10.1 Classification of domain-independent techniques for improving reliability and reducing risk by exploiting asymmetry.

10.2 REDUCING THE LOSSES FROM FAILURES BY ASYMMETRIC STRENGTHENING PROPORTIONAL TO THE COST OF FAILURE

The cost of failure of components is often highly asymmetric. Understanding and strengthening according to the asymmetric cost of failure can significantly reduce the losses from failures of the entire system. The technique of asymmetric strengthening according to the cost of failure of the components allocates more resources towards strengthening components with higher cost of failure.

An example of reducing the losses from failures by asymmetric strengthening proportional to the cost of failure can be given with two very simple systems consisting of only two components, logically arranged in series (Figure 10.2).

For the first system (Figure 10.2a), suppose that component *A1* fails on average twice a year ($f_{A1} = 2$) and its failure is associated with $C_{A1} = 2000$ units of losses. Similarly, component *B1* also fails on average twice a year ($f_{B1} = 2$) and its failure is associated with $C_{B1} = 100$ units of losses. Suppose now that for an alternative system consisting of the same type of components *A* and *B* (Figure 10.2b), the losses associated with failure of the separate components are the same, but the failure frequencies are different because of an asymmetric strengthening of the components. Resources towards reliability strengthening have been reallocated from component *B* towards component *A*. As a result, the reliability of component $A2$ has been improved and the component is now characterized by $f_{A2} = 1$ average number of failures per year while component $B2$ is now characterized by $f_{B2} = 5$ average number of failures per year. Since the first system fails whenever either component *A1* or component *B1* fails, the expected (average) losses from failures $\bar{L}_1$ for the system in Figure 10.2a are

$$\bar{L}_1 = f_{A1}C_{A1} + f_{B1}C_{B1} = 2\times 2000 + 2\times 100 = 4200$$

while for the system in Figure 10.2b, the expected losses from failures $\bar{L}_2$ are

$$\bar{L}_2 = f_{A2}C_{A2} + f_{B2}C_{B2} = 1\times 2000 + 5\times 100 = 2500$$

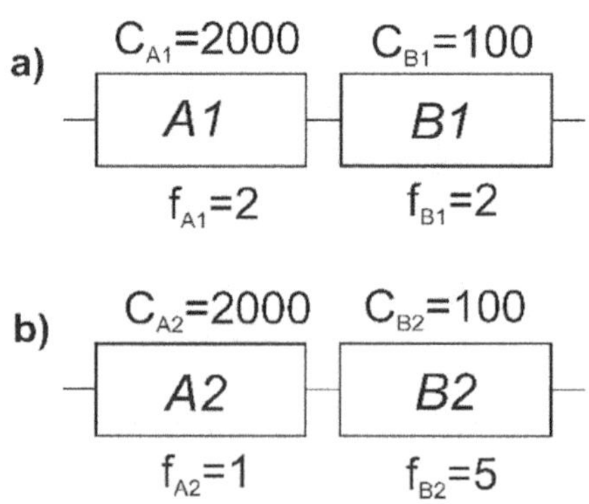

FIGURE 10.2 (a) A system composed of two components, each characterized by the same reliability and (b) a system demonstrating that asymmetric strengthening of the components reduces the losses from failures.

As can be verified, the asymmetrically strengthened system in Figure 10.2b, albeit less reliable than the system in Figure 10.2a, is associated with smaller losses from failures.

This simple example demonstrates that uniformly strengthening components against failure can lead to significant losses. However, in the common case where system failures have varying associated costs, asymmetrically strengthening components, proportional to their failure costs, can significantly reduce the overall losses from system failures.

10.3 IMPROVING RELIABILITY BY ASYMMETRIC STRENGTHENING ACCORDING TO THE ANTICIPATED LOADS

Loading is often asymmetric and understanding and reinforcing for asymmetric loading can significantly improve the reliability of various systems. By anticipating and accommodating asymmetric forces, engineers can ensure that mechanical components and structures perform reliably under real-world conditions.

The technique of asymmetric strengthening according to the anticipated loads also ensures that the structure can handle the highest expected loads without over-engineering other parts, thereby improving reliability while optimizing material use.

A common example of improving reliability by asymmetric strengthening according to the anticipated loads is the asymmetric strengthening of the outer side of car's tire to improve resistance to impact with a curb.

Another example is the joint design tailored to specific loading conditions or operational demands. In many joint configurations, stresses are not uniformly distributed; certain areas experience higher stresses than others. An asymmetric design strengthens the joint precisely where high stresses occur (Coria et al., 2020). This approach is particularly critical under cyclic load conditions, where fatigue failures commonly initiate in high-stress regions. When a joint primarily experiences loading in a specific direction, asymmetric reinforcement can improve its load-bearing capability in that direction. By adding reinforcing material only where it is critically needed and reducing it where it is less essential, asymmetric strengthening in joint designs can save material and reduce weight. This results in cost savings, reduced inertia in dynamic systems, and improved fuel efficiency in vehicles.

Roads are designed with varying pavement thickness, typically thicker in areas expected to bear higher loads, such as near intersections or curves. This asymmetric strengthening extends the pavement's lifespan and enhances its reliability.

One more example of improving reliability by asymmetric strengthening according to the anticipated loads is using stiffer springs or stronger materials on the side of the vehicle that typically bears more load. This measure reduces wear and improves vehicle handling.

Offshore wind turbines subject to uneven forces from wind, waves, and currents is yet another example. The foundations of wind turbines are often designed asymmetrically, with additional reinforcement on the windward side to handle the higher loads, thereby enhancing the reliability and stability of the entire structure.

10.4 IMPROVING RELIABILITY AND REDUCING RISK BY EXPLOITING AN ASYMMETRIC RESPONSE FROM INVERSION

Exploiting asymmetric response from inversion is an effective domain-independent reliability-improvement technique that leads to simple, low-cost solutions. Following this technique, particular features of the system or process are inverted so that the same set of required functions is delivered by the system, but failure modes do not appear or, if they do appear, their impact is minimized. The inversion technique relies on exploiting the inherent asymmetry found in engineering systems and processes.

The asymmetric and dual nature of many characteristics and properties often permits reducing vulnerability and eliminating failure modes by *shifting towards the other extreme of the characteristic/property while preserving the required function.* Thus, shifting from forward to backward motion, motion to rest, and rest to motion, often creates a response that eliminates critical failure modes.

Inversion can be done by inverting relative position, orientation, motion, functions, features, and properties.

An example of an asymmetric response achieved by inverting position, that causes a specific failure mode to disappear, can be given with the cover of a container under pressure in Figure 8.3 from Chapter 8. By inverting the position of the cover from outside to inside, the failure mode 'leakage from the cover seals' is eliminated because the internal pressure provides an asymmetric response helping to strengthen the cover seal.

10.5 EXPLOITING COMPONENTS AND SYSTEMS WITH ASYMMETRIC RESPONSE

This technique leverages asymmetric response to favourably bias the system's response in a way that enhances reliability. Diodes are typical components with passive asymmetric response. They allow current to flow in only one direction, preventing reverse currents that could damage circuits.

Check valves are another example of a component with passive asymmetric response. They allow fluid to flow in one direction only, preventing backflow that could lead to system failure or contamination.

Torsion bars in vehicle suspensions that allow controlled twisting to absorb shocks in one direction while resisting excessive twisting in the opposite direction is yet another example of a component with passive asymmetric response.

Systems with active asymmetric response dynamically adjust the system's behaviour based on its operating conditions, ensuring that the system remains reliable under a wide range of scenarios. As a rule, such systems utilize sensors and control algorithms. The active asymmetric response allows a system to adapt its behaviour in real-time to counteract potential sources of unreliability. By making real-time adjustments, the system can remain reliable under a wide range of operating conditions.

Active asymmetric response is based on sensors and feedback, for example, by using sensors to monitor the system's operating conditions and performance. If a potential source of unreliability is detected, the system adapts its behaviour to counteract it by making adjustments that are inherently asymmetric. For example, if one part of a system is overloaded, asymmetric movement of a counterweight is activated so that the overload is counterbalanced (Ciupitu, 2018; Zhen et al., 2017).

10.6 EXPLOITING ASYMMETRIC GEOMETRY

Asymmetric geometry is a powerful design principle in engineering and manufacturing, often employed as a *poka-yoke* (error-proofing) technique to ensure correct assembly and prevent incorrect orientation of components. For example, some connectors are intentionally shaped like trapezoids rather than rectangles to prevent rotation and ensure proper alignment with their sockets (Figure 10.3a). Similarly, the corner of a SIM card is cut to ensure that the card is inserted in the right orientation. This prevents the user from mistakenly inserting the SIM card upside down or backward, which could potentially damage the card or the device.

This principle is also applied in *keyed shafts and couplings*, where a flat edge or notch on the shaft fits precisely into a corresponding slot in a gear or pulley. This asymmetric interface ensures correct angular positioning and prevents rotational slippage.

In *aircraft engine mounts*, asymmetric placement of mounting holes is used to prevent the installation of an engine on the wrong side of the aircraft, a critical safety measure in twin-engine planes. By physically preventing incorrect assembly, asymmetric geometry enhances both functionality and safety across various engineering products.

Asymmetric geometry often translates into asymmetry in stiffness. The asymmetry in stiffness ensures that when external forces are applied, the component or system naturally flexes or deforms in a manner that optimizes load distribution and minimizes stress intensification. When a force is applied, the component will naturally bend or shift in a manner that channels the load through the stiffer regions while allowing more flexible areas to deform without breaking.

Beams with an asymmetric cross-section can offer tailored deflection properties, ensuring they flex or bear load more effectively in specific directions or under unique loading conditions. Thus, a tapered (asymmetric) cantilevered beam, despite using the same volume of material, possesses a superior load-bearing capacity compared to its straight-beam counterpart.

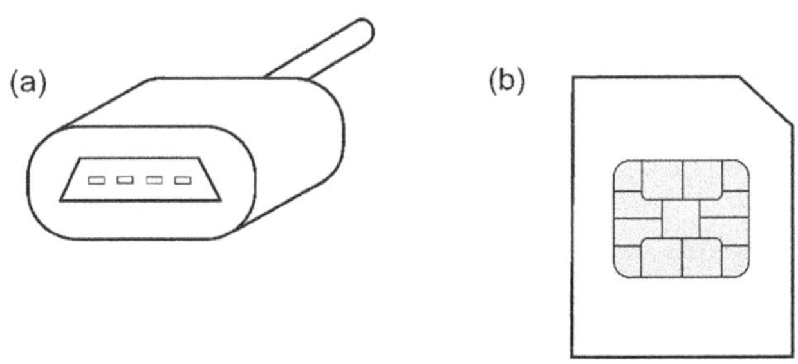

FIGURE 10.3 Asymmetric geometry ensuring a poka-yoke design (a) connector and (b) a SIM card.

Aircraft wings are designed with varying thickness in critical areas to handle asymmetric loads thereby improving the wings' fatigue life and overall structural integrity.

Asymmetric geometry can also be employed in fluid systems. Asymmetrically curved channels can help in reducing cavitation, turbulence, or promoting laminar flow, thereby extending the life of the system.

Asymmetric geometry is used in the design of impeller blades. For pumps or fans, blades with an asymmetric profile can optimize fluid or air movement, increase efficiency and reduce wear.

In load-supporting beams, having asymmetric geometry for the web and flange (e.g., the T-beams), can increase the strength along anticipated load directions.

10.7 EXPLOITING ASYMMETRIC MATERIAL PROPERTIES

Exploiting asymmetry in properties can provide improved mechanical reliability in many engineering applications.

Using hybrid materials is a typical example of combining materials with different properties in an asymmetric manner, where the high-performance material is placed in regions of high stress or wear, and the low-performance material is used in less critical areas. In components such as turbine blades, the heat-resistant ceramic is used in high-temperature zones, while the alloy part is used in low-temperature zones.

Next, using different materials in asymmetric configurations at joints or bonds often enhances reliability in specific directions. This is achieved by bonding materials with complementary properties, like combining a tough material with a hard, wear-resistant one. In the aerospace and automotive industries, a high-strength alloy is often used for the structural part of a frame, while bonding it with a less stiff material at the interface to absorb vibrations.

Asymmetric properties created by heat treatments are often used to modify material properties, such as hardness, toughness, or ductility. Commonly, the tooth flanks of gears, which experience a high rate of wear, undergo such heat treatment to increase hardness, while the microstructure in the central zones of the gears is left unchanged to preserve toughness. Asymmetric material properties that strengthen the surface and increase fatigue resistance in components can also be achieved through localized laser treatment or electron beam processing.

An additional, well-known technique that leverages asymmetric material properties is the use of functionally graded materials. These are composite materials where the composition or the microstructure is intentionally varied in space to create a gradient in specific properties. By creating a material with varying hardness, an incompatible combination of properties can be achieved that provides enhanced reliability: for example, a hard exterior for wear resistance and a ductile interior for impact resistance.

Functionally graded materials can be designed to reduce the thermal stresses that develop at the interface of two different materials. By grading the material, the thermal expansion mismatch is reduced, and the thermal stress is distributed over a larger volume. In jet engine applications, functionally graded materials can serve as effective thermal barriers between the hot gases and the underlying metal components. The gradient in thermal conductivity prevents the metal components from overheating.

Ceramic-alloy functionally graded materials which combine a hard, wear-resistant, and thermal-resistant surface with a tough subsurface inhibiting crack propagation, exemplify this principle (Figure 10.4).

The property gradients in some functionally graded materials enable parts of the structure to be biocompatible while retaining substantial mechanical strength elsewhere, making them ideal for dental and orthopaedic applications.

Finally, functionally graded coatings offer improved reliability compared to conventional coatings by providing stronger bonding and reducing thermal and residual stresses.

Key fabrication techniques for producing thin and bulk functionally graded materials with tailored gradients include vapour deposition, powder metallurgy, centrifugal casting, and solid freeform fabrication (Mahamood et al., 2012).

10.8 EXPLOITING ASYMMETRIC LOADING

Loading can be designed so that wear is deliberately uneven across components. This decreases the risk of failure by ensuring that wear is deflected from components with large consequences of failure towards easily replaceable components associated with small consequences of failure.

Loads can also be applied asymmetrically across different components or systems based on real-time performance metrics. For example, high-performance vehicles experience dynamic loads, especially during cornering. Asymmetric suspension setups deliberately distribute loads unevenly to improve traction and handling in specific conditions, which enhances the vehicle's reliability by preventing overloading and failure of suspension components.

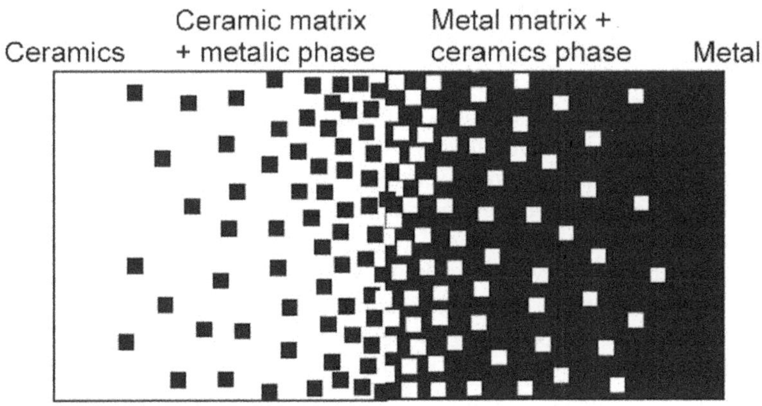

FIGURE 10.4 Ceramic alloy as a functionally graded material.

10.9 EXPLOITING ASYMMETRIC POSITIONING

Asymmetrically positioned elements building a system can prevent the entire system from reaching a resonance frequency, which can be destructive. Resonance points in a system are identified, and elements are positioned asymmetrically to avoid resonance.

Asymmetric positioning of heavier components in a system can be done for attaining optimal centre of gravity, thereby improving stability and performance, especially in mobile systems.

In mechanical assemblies, fasteners (like bolts or rivets) may be placed asymmetrically to account for uneven loading. For instance, in a structure where one side is expected to bear more load, additional or stronger fasteners might be used on that side. This approach enhances the reliability of the joint by preventing failure under asymmetric overload.

Asymmetric blade positioning has been proposed as an effective strategy to mitigate resonance phenomena and thereby reduce helicopter rotor noise (Edwards and Cox, 2002). Rotor noise primarily originates from the main rotor system, where equally spaced blades generate periodic aerodynamic loading. This periodicity induces constructive interference of the resulting sound waves, amplifying overall noise levels. By arranging the rotor blades asymmetrically, i.e., at unequal angular intervals, the periodicity of the excitation is disrupted, resonance effects are suppressed, and the radiated noise is significantly reduced.

10.10 EXPLOITING ASYMMETRIC WEAR AND THE ASYMMETRY IN AGEING

Asymmetric wear of brake pads equipped with wear indicators offers a practical example of intentionally engineered asymmetric degradation that enhances both reliability and safety. Brake pads function by pressing against a rotating brake disc to slow or stop a vehicle, gradually wearing down due to friction. If the pad material wears away completely, the exposed metal backing plate can contact the disc, potentially causing severe damage or brake failure. To prevent severe damage of the brake disks, wear indicators are deliberately designed to wear asymmetrically – faster than the surrounding friction material, thus creating an early warning system. These indicators can take the form of a physical notch or groove that disappears as wear progresses. This intentional design provides drivers with a clear visual signal to replace the brake pads before they become dangerously worn.

Components are made to match their expected lifespan. For example, components that are more wear-resistant or fatigue-resistant are used in areas subjected to high stress. Durable and expensive components are used in high-wear areas and cheaper, less durable components are used in low-stress areas.

Often, it is beneficial to use components with varying lifespans and replacing shorter-lived parts more frequently to protect longer-lasting and more expensive parts. Regular replacement of components with short lifespans that experience wear or degradation over time can extend the lifespan of more expensive critical parts and ensure that the system continues to meet performance standards without requiring a full replacement.

This technique is widely used in industries like consumer electronics, automotive, and heavy machinery, where parts like filters, gaskets, or seals are designed for easy replacement without significant system disruption.

10.11 EXPLOITING ASYMMETRIC PAYOFFS

This technique is about designing strategies where the potential downside is limited while the upside remains significant. Asymmetry is exploited by ensuring that while the exposure to negative outcomes is limited, the potential for positive outcomes remains unchanged or is amplified.

An example can be given with portfolio diversification with asymmetric payoffs. Such portfolios include assets with asymmetric payoffs, such as options, where the potential upside is large while the downside is limited to the initial investment.

Another relevant example can be given with setting a stop-loss in trading. It ensures that a stock is automatically sold before losses become too large, thereby preserving potential gains and limiting potential losses.

Paying for insurance is yet another example: it has a small, predictable cost, but it protects against catastrophic financial losses. A similar example can be given with investing in monitoring equipment. It is associated with a modest initial investment but can prevent catastrophic system failures associated with huge financial losses.

Exploiting asymmetric payoffs is a powerful strategy with wide-ranging applications beyond engineering. Consider health management, for instance. Imagine a critically ill patient faced with four life-saving treatments: A, B, C, and D, each carrying the risk of severe side effects that could cause long-term harm. Despite these side-effect risks, the immediate need to preserve life justifies proceeding with treatments A, B, and C. Suppose that after completing these, the patient's condition stabilizes, and their life is no longer in imminent danger. At this point, the decision to proceed with treatment D introduces another asymmetric payoff: the potential harm from side effects now outweighs the marginal benefit of further improving an already stable condition. In such cases, recognizing and leveraging the imbalance between risk and reward becomes crucial for optimal decision-making.

The next two sections focus on two new powerful low-cost techniques for enhancing reliability and reducing risk by exploiting asymmetry: 'asymmetric arrangement of interchangeable redundancies' and 'exploiting asymmetric output characteristics'.

10.12 EXPLOITING ASYMMETRIC OUTPUT CHARACTERISTICS OF SYSTEMS AND PROCESSES

The idea behind this technique is to use components or processes whose response curves naturally suppress harmful deviations. Such components absorb or mitigate deviations that would otherwise compromise robustness. A typical example is Zener diodes, which clamp voltage spikes above a certain threshold while allowing normal operation below it.

Another example is an output characteristic that exhibits a saturation region. In the steep region of the output characteristic, a variation Δx in the input parameter

x of a given magnitude causes a large variation Δy in the output (response), which is undesirable in many applications. In contrast, in the saturated region of the output characteristic, the same variation Δx in the input parameter x results in only a small variation Δy in the output (Figure 10.5). If the process is conducted in such a way that the input is within the saturated region of the characteristic, inevitable variations in the input parameter will not lead to large variations in the output.

10.13 IMPROVING SYSTEM RELIABILITY BY ASYMMETRIC ARRANGEMENT OF REDUNDANCIES

Chapter 9 provides a detailed discussion of this technique, and we will not repeat it here. In brief, for series-parallel systems, the asymmetric arrangement of interchangeable redundancies consistently results in higher reliability compared to a symmetric arrangement, regardless of the failure probabilities. Even when the failure probabilities of components and their reliability rankings are unknown, a series-parallel system with redundant components can have its reliability significantly enhanced by adopting an asymmetric arrangement of the redundancies.

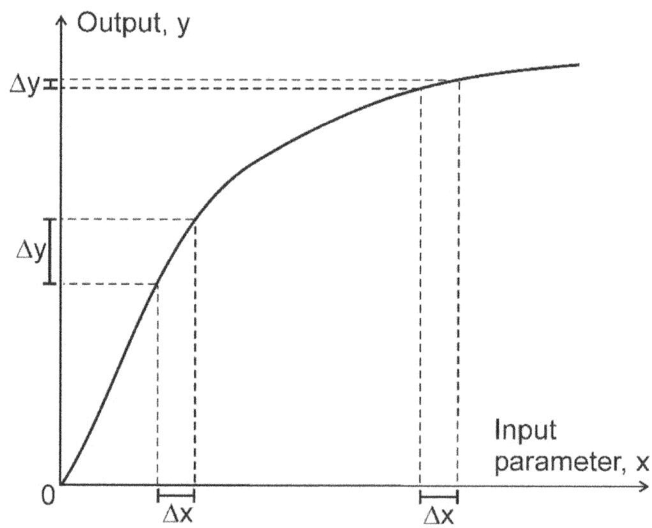

FIGURE 10.5 An asymmetric output characteristic, characterized by a saturation region.

11 Improving Reliability by Increasing the Level of Balancing

11.1 CLASSIFICATION OF TECHNIQUES FOR INCREASING THE LEVEL OF BALANCING

The underlying idea of this method is *improving reliability by ensuring increased balance of loads, properties, and reliability-critical parameters over the different parts of the system.*

In well-balanced systems, load is distributed more uniformly across components, and as a result, the loading stresses are smaller. Distilling numerous examples of improving reliability by increasing the level of balancing and organizing the information yielded the following major techniques (Figure 11.1):

- *Ensuring a more uniform distribution of loads;*
- *Altering the transmission paths of forces;*
- *Ensuring a more uniform distribution of internal stresses;*
- *Ensuring a more uniform distribution of material properties;*
- *Improving the stability of a system;*
- *Reducing the magnitudes of inertia forces;*
- *Capturing a compensating factor;*
- *Reducing the variability of reliability-critical parameters.*

Although in several practical solutions reliability has been effectively improved by increasing the level of balancing, improving reliability by increasing the level of balancing has not yet been recognized as a generic reliability improvement method in reliability engineering and engineering design. No analysis of the mechanisms and techniques through which increasing the level of balancing improves reliability has been presented in standard texts related to reliability engineering and engineering design. The outlined gap in the relevant published literature prompted a recent publication (Todinov, 2022a) covering this generic method for improving reliability and a classification of relevant techniques (Figure 11.1).

DOI: 10.1201/9781003679301-11

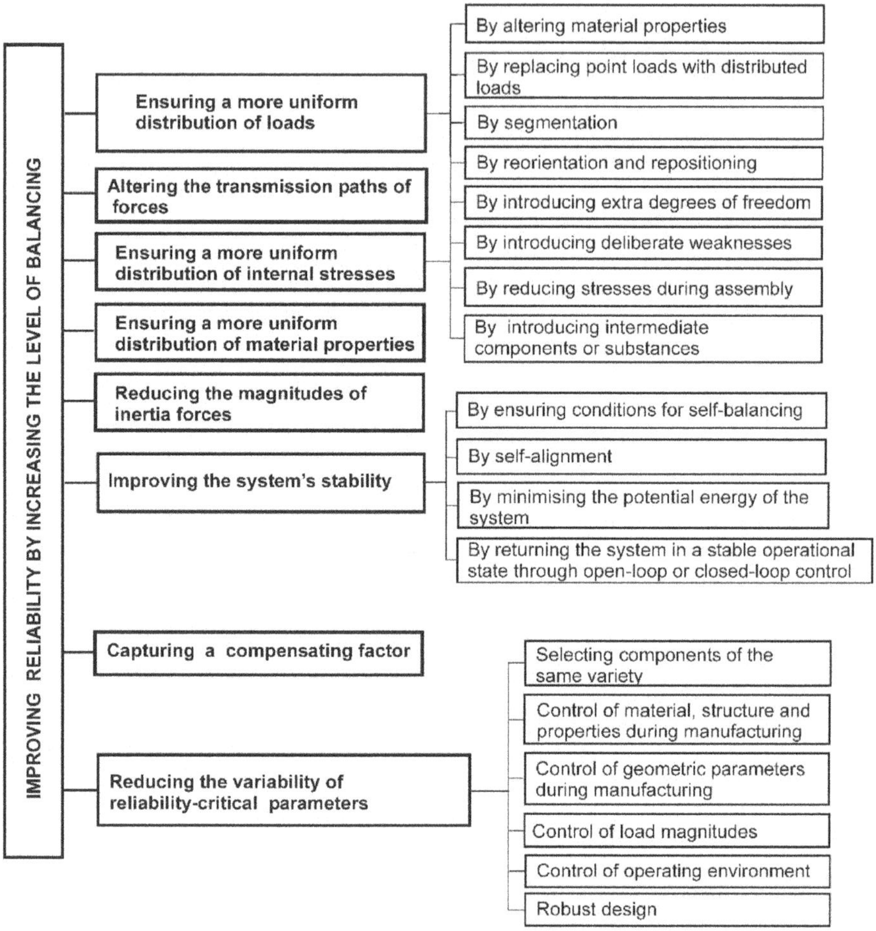

FIGURE 11.1 Classification of techniques related to improving reliability by increasing the level of balancing.

11.2 APPLICATION EXAMPLES RELATED TO DIFFERENT TECHNIQUES FOR INCREASING THE LEVEL OF BALANCING

A more uniform load distribution leads to significantly reduced contact stresses and wear.

In mechanical systems, rather than concentrating most of the load on a single component (e.g., a screw), which can lead to excessive stress and failure, the load can be distributed among multiple components. The load sharing reduces the force experienced by each individual component and enhances overall system reliability. Imbalanced electrical loads, for example, frequently cause excessive heat generation and reduced reliability of electrical devices. An adequate assignment of the parameter values of the components in electrical circuits improves the level of balancing of electrical loads and reliability. Imbalanced input voltage, for example, could lead to imbalanced input power of the modules in electronic circuits which creates a possibility of exceeding voltage rating of some electronic components and system failure. This can be prevented by increasing the level of balancing through a voltage-sharing balance circuit (Kroics et al., 2016).

Imbalanced flow of data in computer networks frequently results in delays and lost packets of information. Proper management of the data flow by employing prioritization algorithms improves the level of balancing of the data flow and improves the quality of service of the network.

Self-balancing of transportation networks involving autonomous vehicles could make full use of the intelligent characteristics of self-driving vehicles and improve the utilization rate of the vehicles in the system. This improves the distribution of transportation resources in the road network and the efficiency of vehicle service (Li et al., 2021).

Point loading often causes excessive deflection of the load-carrying surfaces, excessive stress, and failure. Unless the load-carrying surface has been specifically designed to accommodate point loads, the point loads often cause excessive deformation and failure. Increasing the level of balancing of external loads by distributing them uniformly over the load-carrying surface decreases deflections and stress and reduces the likelihood of failure. Including deliberate weaknesses or extra degrees of freedom, for example, does not permit the formation of peak stresses and contributes to a more uniform distribution of internal stresses. Such is, for example, the case where extra degrees of freedom are introduced in statically indeterminate structures to turn them into statically determinate structures. The result is a more uniform distribution of the internal stresses and reduced peak stress magnitudes.

The next case study demonstrates increasing the level of balancing of external loads by distributing them uniformly over the load-carrying surface.

11.3 CASE STUDY: IMPROVING THE RELIABILITY OF DRIVING PILES INTO THE GROUND BY INCREASING THE LEVEL OF BALANCING

In what follows, an example will be given related to improving reliability by increasing the level of balancing. The example is related to concrete piles driven into the ground (Orloff, 2006) (Figure 11.2).

Problem

In Figure 11.2, driving pile (1) into the ground (3) causes damage to the top end of the pile because of uneven distribution of the contact pressure from the hammer (2). Part of the top end of the concrete pile is overstressed while other parts are underloaded and the result is excessive stress on part of the cross-section which leads to cracks and damage of the concrete pile.

Going through the classification of techniques characterizing the method of increasing the level of balancing suggests that the most appropriate technique is '*ensuring a more uniform distribution of the load by introducing intermediate components or substances*'.

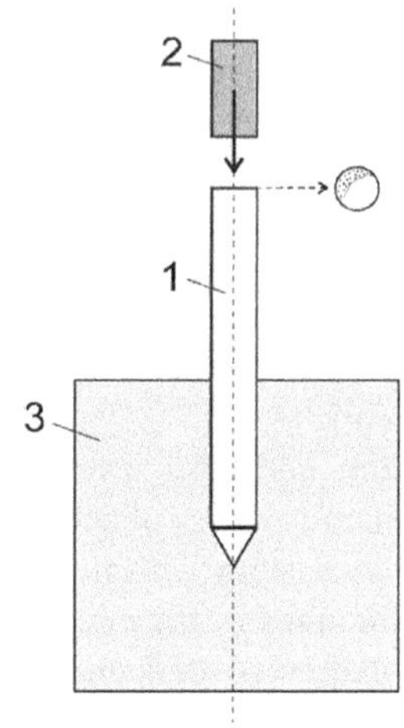

FIGURE 11.2 A process of driving a pile into the ground with a failure mode 'damaging the top of the pile'.

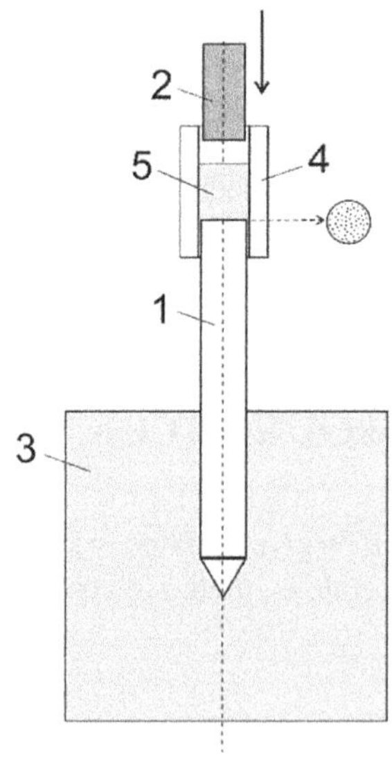

FIGURE 11.3 Eliminating the failure mode 'damaging the top of a pile driven into the ground' by increasing the level of balancing through an intermediate component.

Introducing a cheap intermediate component '5' between the hammer '2' and the pile '1' by using the sleeve '4' (Figure 11.3) improves the uniformity of the load distribution over the top end of the pile (Orloff, 2006). This results in reduced contact stress and reduced likelihood of damaging the pile. The intermediate component can simply be sand or other cheap material.

11.4 REDUCING THE MAGNITUDE OF THE INERTIAL FORCES BY INCREASING THE LEVEL OF BALANCING

Balancing is a fundamental practice in mechanical engineering that reduces the magnitude of inertia forces in rotating components. It plays a critical role in enhancing mechanical integrity, increasing efficiency, and ensuring operational safety. Whether achieved through static or dynamic methods, proper balancing minimizes mechanical stress and is a cornerstone of reliable machine design.

The goal of balancing is to reduce or eliminate the net inertia forces and moments caused by uneven mass distribution in a rotating system. This is usually accomplished by aligning the centre of mass with the axis of rotation or by strategically adding counterweights. Balancing of rotating components can be *static* and *dynamic* balancing (Uicker et al., 2003).

Static balancing ensures that the centre of mass is aligned with the axis of rotation. In this condition, no net force produces a moment about the axis of rotation when the component is at rest. This type of balancing is particularly suitable for components that rotate at low speeds, such as flywheels.

Dynamic balancing, on the other hand, accounts for both unbalanced forces and moments that arise during rotation. It is critical for high-speed rotating components like turbines and crankshafts. In such cases, unbalanced centrifugal forces generate unbalanced couples, which can be neutralized by introducing a counteracting couple of equal magnitude and opposite direction.

There are numerous benefits from balancing rotating components:

- A balanced rotor generates uniform radial acceleration, preventing the buildup of forces in specific directions. This reduces oscillations and mechanical stress on bearings and supports.
- Balanced systems have extended lifespan. By reducing dynamic loads on bearings, shafts, and housings, balancing slows down fatigue and wear, thereby increasing service life. In high-speed applications such as aircraft engines, centrifuges, and turbines, balancing helps prevent catastrophic failures due to excessive dynamic forces.
- Unbalanced forces lead to vibrations that can damage machinery and reduce operational lifespan. Balancing significantly reduces vibrations, and provides conditions for quieter and smoother operation.
- Balanced systems have improved mechanical efficiency. They waste less energy countering unbalanced motion. This enhances power transmission and fuel efficiency, particularly in engines and drive systems.

Balancing is performed by adding or removing mass, through drilling, grinding, or adding counterweights, either manually or using balancing machines. These machines can be static (for single-plane corrections) or dynamic (for multi-plane corrections).

A typical example is the crankshaft balancing in engines. In internal combustion engines, rotating and reciprocating components such as crankshafts and pistons create significant inertial forces. To balance these forces, counterweights are added to the crankshaft. This compensates for both rotating and reciprocating masses, resulting in smoother engine operation, reduced wear, and greater longevity.

11.5 INCREASING THE LEVEL OF BALANCING BY ALTERING THE TRANSMISSION PATHS OF FORCES

Altering the transmission paths of forces contributes significantly to increasing the level of balancing and improving the reliability of assemblies and structures. By re-engineering how forces are transmitted, systems can become more robust, longer-lasting, and safer to operate.

As a rule, short and direct transmission paths are associated with a minimal amount of deformation and a higher level of balance. Avoiding eccentric axial loading of columns by applying force through the centroid of the cross-section eliminates the bending moment associated with eccentric loading and leads to improved stability of the column (Hearn, 1985).

External forces applied through the centres of symmetry of sections and along axes of rotation are associated with a higher level of balance due to the absence of bending moments. A higher level of balance is also achieved if the transmission path of the external force lies within the cone of friction.

Re-routing forces through stronger or more durable paths can be accomplished by designing force paths to pass through stronger structural members rather than through joints or areas prone to fatigue. For example, axial loads can be redirected through a central housing instead of bolted flanges.

Decoupling components is another method of altering the transmission paths of forces. This includes the use of flexible couplings, isolation mounts, or intermediate elements to isolate sensitive components from high-force transmission. Decoupling prevents force transmission directly into vulnerable areas (e.g., isolating electronic modules from engine vibrations).

Altering the transmission paths can also be achieved by modifying geometries to redistribute loads. Component shapes (e.g., fillets, ribs, arcs) are adjusted to spread forces more evenly. This results in reduced local stress concentrations, which are common points of failure initiation.

11.6 INCREASING THE LEVEL OF BALANCING BY MORE UNIFORM DISTRIBUTION OF INTERNAL STRESS

In designing components, a more uniform distribution of internal stress can be achieved by avoiding stress concentrators (e.g., sharp corners or sudden changes in cross-section). These lead to elevated local levels of stress that can easily exceed the strength of the material even at relatively low nominal internal stress away from stress concentrators. Instead, use smooth geometrical transitions, fillets, or radii at corners to reduce stress concentrations. Finite Element Analysis can be used to identify stress concentrations and adjust the design accordingly.

To avoid stress concentrators, uniform cross-sections should be used where possible. Using stress relief features (slots, holes, or grooves) can also be used to relieve stress concentrations. Strategic placement of these can 'smooth out' areas of high stress.

Imbalanced internal stresses in loaded components, however, can be present even in the absence of stress concentrators. In the design of composite rods, composite beams, and composite torsion bars, the limit state corresponding to the onset of failure (e.g., yielding) in the constituent parts can occur at significantly different levels of the loading force. This leads to an unbalanced design, where parts made of a particular material are overdesigned while parts made of another material are designed with insufficient strength. Commonly, unbalanced designs of composite components are characterized by significantly different factors of safety related to the constituent parts. In many cases, such designs can be balanced by altering the geometry of the constituent parts made of different materials in such a way that the limit state associated with the different parts occurs at close magnitudes of the loading force. The distribution of internal stresses can also be made more uniform by segmenting external loads or distributing the external loads instead of concentrating them over a small spot (Todinov, 2019a).

Material selection and treatment are also essential for the uniform distribution of internal stresses. Materials with uniform microstructures and high ductility can be chosen to better distribute stresses. In addition, heat treatment (e.g., annealing) can be applied to reduce unfavourable tensile residual stresses.

Pre-stressing of components can also be used to counteract expected service loads. This method intentionally introduces beneficial internal stresses to increase the level of balancing and reduce the magnitude of the net stresses.

11.7 USING SEGMENTATION AND INTERMEDIATE COMPONENTS TO IMPROVE THE LEVEL OF BALANCING

As a rule, increased segmentation of external loads increases the level of balancing and improves the load-carrying capacity of assemblies.

Thus, for high-torque applications, planetary type of gearbox is often preferred for transmitting torque because of the increased number of contact points (three instead of one) (Bakshi et al., 2014). This type of design reduces the contact stress on the gear teeth and enhances durability.

Similarly, transitioning from a key connection in shaft/hub assembly to a spline connection is beneficial because torque is transmitted through multiple elements (instead of a single element) and, as a result, the contact stresses are reduced.

Often, the contact of two components with high hardness does not permit a uniform distribution of the contact stresses. As a result, some zones from the contacting surfaces are characterized by high contact stress intensity while for others, the stress intensity is low. Using elastic components as mediators (e.g., rubber or other appropriate material) improves the distribution of contact stress and reduces its magnitude. Increasing the level of balancing through intermediate components provides protection against elevated stresses from shock loading.

An example of alteration of geometry to improve the level of balancing can be given with the use of different pitch values to ensure uniform thread load distribution in bolted joints (Coria et al., 2020). More uniform distribution of the stresses along the thread of a bolt connection is also ensured by selecting the material of the nut to be of smaller stiffness than the material of the bolt (Coria et al., 2020).

As noted, segmentation is beneficial in reducing the stresses in loaded assemblies; however, it is associated with a downside. Often, the loads are not distributed uniformly across the separate segments. This is a technical contradiction associated with segmentation, and possible ways of resolving it and improving the load distribution over the separate segments is the alteration of material properties, geometry, or structure (Todinov, 2022a).

11.8 INCREASING THE LEVEL OF BALANCING THROUGH CONTINUITY AND HOMOGENEITY

It is a well-documented fact that fatigue life depends strongly on the amplitude of the loading stress. According to the Paris-Erdogan law (Paris and Erdogan, 1963), the rate r of fatigue crack propagation is given by

$$r = C(\Delta K)^m$$

where ΔK is the stress intensity factor range while C and m are constants. The stress intensity factor range is directly proportional to the stress range $\Delta\sigma$ from loading.

Avoiding discontinuous operation (start-stop regimes) of pressure vessels and pipelines significantly reduces the range of the thermal and loading stresses and greatly enhances the fatigue life of these components.

Ensuring continuity in welding and hot rolling manufacturing operations avoids defects associated with the interruption of the operation and improves the quality of the final product. Eliminating transitional regimes in electrical circuits with large inductors, caused by unnecessary start-stop regimes, eliminates excessive voltages and the risk of damaging components in the circuit. Shifting from a linear reversed motion to a rotation eliminates large inertia forces associated with reversing the motion.

Homogeneity in material properties is an important factor for increasing reliability by increasing the level of balancing. A uniform heating and cooling of a component with non-uniform thermal properties in the core and at the surface layers causes the appearance of a significant thermal gradient and thermal stresses. Their magnitude can be so large that they often lead to the fracture of components with large sections. In contrast, uniform thermal properties of heat-treated components guarantee a significant reduction of the internal thermal stresses upon heating and cooling.

Furthermore, eliminating the anisotropy of the microstructure in loaded components eliminates the anisotropy of strength and creates better conditions to resist forces acting in different directions. Guaranteeing a structure with uniform grain size in an alloy, eliminates excessive variation of the fracture toughness and creates better resistance to fatigue crack propagation.

11.9 INCREASING THE LEVEL OF BALANCING BY IMPROVING THE SYSTEM'S STABILITY

Stability is a property of vital importance to engineering systems. For a stable system, a bounded input produces a bounded output, and the system returns in equilibrium state once the external disturbances have been removed. The system's stability can be improved significantly by ensuring conditions for self-balancing.

11.9.1 Improving System's Stability by Self-Balancing

Unwanted forces create excessive internal stresses in components and assemblies and require extra material or reinforced material to resist them. As a result, the weight and cost of assemblies are increased. The extra loads also create extra friction forces, premature wear, fatigue degradation, and failure.

The need for extra material, reinforced material, and heavier construction can be avoided if conditions for self-balancing are created. As a rule, ensuring conditions for self-balancing in a system improves the system's reliability (Meraz et al., 2005).

Self-balancing can be achieved by (i) full or partial compensation of forces or negative factors or by (ii) stabilizing the system through open-loop or closed-loop control.

Self-balancing through symmetrical design can be used to eliminate undesirable forces. Such an example has been discussed in (Matthews, 1998), (Figure 11.4).

The axial force from the turbine '1' requires a thrust bearing '2'. If two symmetrical turbines are mounted on the common shaft (Figure 11.4b), the axial force from each turbine will be compensated by an equal and opposite axial force from the other turbine. As a result, the thrust bearing is no longer needed; degradation caused by wear of the thrust bearing is avoided and the reliability of the assembly is improved.

Herringbone gears also exhibit a similar form of self-balancing, eliminating the need for thrust bearings required in single-helical gears.

Another example of the self-balancing method achieved through symmetrical design can be found in symmetrical epicyclic gear mechanisms, where the radial forces acting on the central shaft are self-balanced.

Increasing the level of balancing can be done by counter-balancing mechanisms designed to take the effort out of moving and supporting heavy loads, lifting loads, and positioning manipulator arms. The result is reduced magnitudes of the required activating forces for maintaining equilibrium states, improved positioning precision, and reduced stresses. Thus, using counterweights in robots avoids the need for

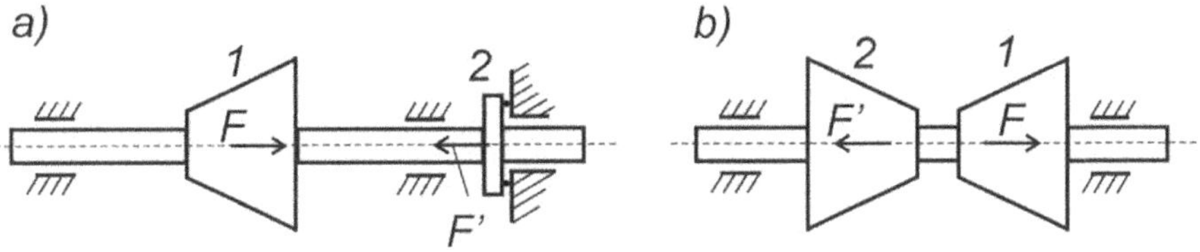

FIGURE 11.4 (a) Unbalanced and (b) self-balancing design.

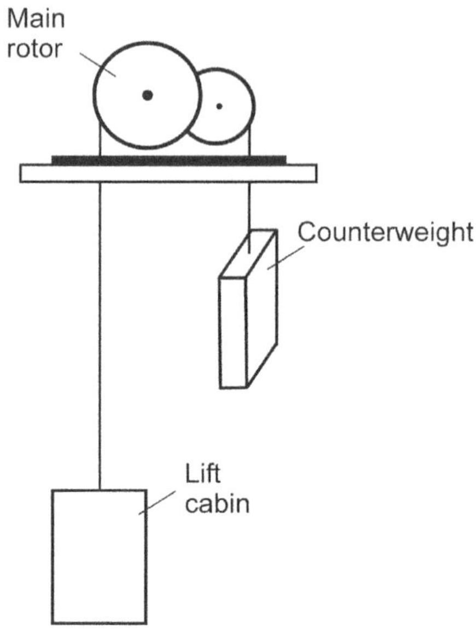

FIGURE 11.5 Lift cabin with a counterweight.

powerful actuators to overcome static loads. For such counter-weight systems, the required motion can be attained by a low-power actuator whose energy consumption is necessary only for overcoming the friction forces (Ciupitu, 2018).

Counterweights in lift carriages (Figure 11.5) decrease the amount of power required from the motor. With counterweights, the required power is much smaller than the power needed to lift the weight of the carriage in the absence of counterweights, which extends the reliability of the motor and its life.

Finally, self-balancing mechanisms can be implemented to reduce the required actuation force, especially when a rapid response is needed. In the configuration shown in Figure 11.6a, the pipe is subjected to a pressure p. To quickly release this pressure, an actuation force F must be applied to the valve. This force must overcome both the spring resistance and the pressure-induced force C (see Figure 11.6a).

In contrast, the self-balancing design shown in Figure 11.6b includes a counterbalancing force D. In this configuration, the actuation force F only needs to overcome the spring resistance and a small net force. As a result, the actuation force required for the design in Figure 11.6b is significantly smaller than the actuation force for the design in Figure 11.6a, enabling more reliable and rapid pressure release.

Even under high pressure p, the valve can still be actuated with a relatively small force F. This is because an increase in the pressure p raises both the pressure force C and the counterforce D proportionally, creating a self-balancing effect.

Similarly, as the next case study shows, a counterpressure can also be used for improving the reliability of cast products.

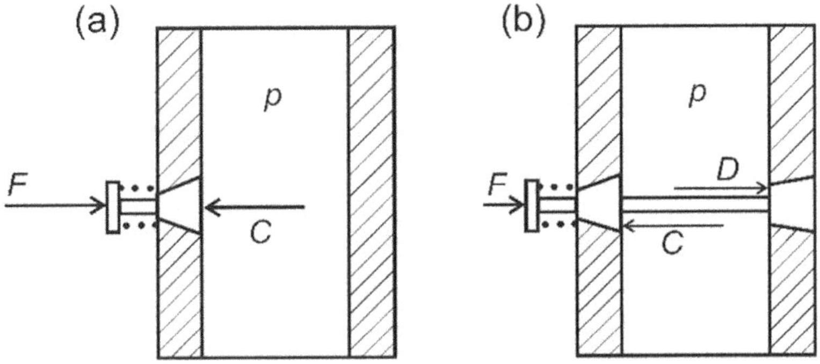

FIGURE 11.6 Self-balancing by creating a counterforce from pressurized fluid.

11.9.2 Case Study: Improving the Reliability of Cast Components by Increasing the Level of Balancing through Counterpressure

Low-pressure casting (Figure 11.7a) is a widely adopted process for producing high-quality, near-net-shape castings, particularly in the automotive and aerospace industries. It is especially well-suited for non-ferrous metals, such as aluminium and magnesium alloys. Despite its advantages, the process presents several limitations and challenges.

While low-pressure casting reduces turbulence and oxidation compared to traditional gravity casting, *porosity* remains a significant concern. A common issue is *gas porosity,* primarily due to the absorption of hydrogen in molten aluminium alloys. As the metal solidifies, the dissolved hydrogen forms bubbles, resulting in voids within the casting. Another prevalent issue is *shrinkage porosity*, which arises from inadequate metal feeding during solidification, especially in regions that are last to solidify.

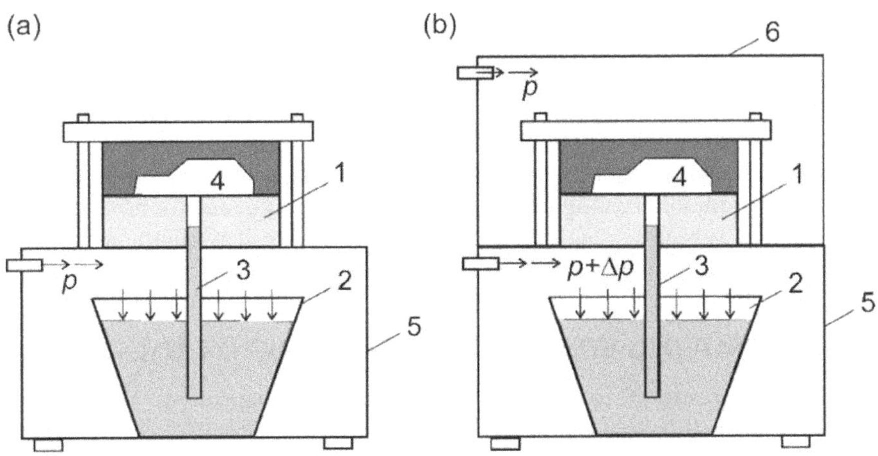

FIGURE 11.7 (a) Low-pressure casting process and (b) counterpressure casting.

Porosity leads to a substantial reduction in fatigue life, tensile strength, and ductility. Pores act as stress concentrators, increasing the risk of crack initiation under cyclic loading. Internal porosity compromises the pressure containment of components such as cylinder heads or hydraulic parts, potentially leading to fluid or gas leakage and part failure. Porosity degrades the surface quality, necessitating additional machining or surface treatments. Pores can trap corrosive substances, promoting pitting corrosion or internal oxidation which is especially critical in marine environment.

Furthermore, the inherently slow filling and controlled solidification of the low-pressure casting limits its productivity, making the process less suitable for high-throughput manufacturing.

These limitations can be effectively addressed by adopting *counter-pressure casting* (Figure 11.7b). By maintaining a higher level of pressure balance and control throughout the filling and solidification stages, counter-pressure casting significantly reduces porosity and improves both mechanical and surface properties, while also enhancing process reliability and productivity.

The primary difference between low-pressure casting and counter-pressure casting lies in how pressure is applied and controlled during the metal-filling process. Both are advanced casting techniques used to produce high-quality components, particularly from aluminium and other non-ferrous metals, but they operate according to different principles.

In the low-pressure casting process (Figure 11.7a), the mould (1) is positioned above a sealed crucible (2) containing molten metal. A low gas pressure is applied to the surface of the molten metal, forcing it upward through a riser tube (3) into the mould cavity (4). This process of controlled and gradual filling reduces turbulence, helping to improve casting quality.

In contrast, the counter-pressure casting process, invented by the Bulgarian engineers Balevski and Dimov (US patent US3693698A [1969]), involves enclosing both the mould (1) and the crucible (2) in separate sealed chambers (5 and 6, respectively). Gas pressure is applied to both chambers, but the pressure $p+\Delta p$ above the molten metal is kept slightly higher than that above the mould cavity. This pressure differential Δp forces the metal into the mould. Crucially, the counterpressure is maintained throughout the solidification phase, enhancing process control.

Compared to low-pressure casting, counter-pressure casting (Figure 11.7b) enables faster but still controlled mould filling. It significantly reduces gas porosity due to the elevated pressure during solidification. The sustained counterpressure also causes microplastic deformation in the solidifying metal, leading to improved mechanical properties. In addition, castings produced by this method generally have smoother surfaces and fewer surface defects. Finally, the sealed system minimizes environmental contamination and enhances workplace safety.

11.10 IMPROVING SYSTEM'S STABILITY BY ACTIVE BALANCING

Multiple components of the same type, working simultaneously, often rely on devices/circuits whose sole purpose is to actively balance the load on the separate components. In these applications, passive balancing is not sufficient to guarantee the required flexibility and resilience and there is a need for active balancing.

An example of a transition from passive to active balancing can be found in crane systems where the counterweight from fixed becomes dynamic, with possibility to be shifted, depending on the operating load (Zhen et al., 2017).

Supercapacitors in particular, are not identical due to manufacturing variations. During the charging of supercapacitors connected in series, at a constant current, the capacitor that first reaches its rated voltage could be damaged if charging continues beyond this point. Supplemental active circuits can be used to monitor and balance the charging process (Ionescu and Drumea, 2019).

Active balancing is also very important for battery cells. The variation in the capacity of battery cells connected in series, causes the voltage of some cells to rise into dangerous levels, which results in a faster degradation of the cells. Active balancing of battery cells is not only important for improving the performance of the battery; it also extends battery life and enhances safety. An active balancing cell equalizer has been proposed in (Jeon et al., 2015).

Active balancing can also be done by open-loop or closed-loop controllers which monitor the controlled process variables and compare their values with specified reference levels. In case of a deviation from these levels, a control action is generated that brings the process variables to match their specified reference values.

11.11 INCREASING THE LEVEL OF BALANCING BY A COMPENSATING FACTOR

In this form of system balancing, a *third factor*, a *compensating element,* is introduced to counteract disturbances and restore equilibrium. Capturing a compensating factor involves leveraging system-inherent responses, where disturbances trigger corrective mechanisms that oppose and mitigate the original perturbation.

For example, *seal designs* should prevent a feedback loop in which frictional heat leads to increased frictional forces. Instead, these designs should ensure that rising temperatures reduce friction, thereby stabilizing the system. Pahl et al. (2007) describe a design involving *taper roller bearings*, where thermal expansion of the shaft acts as a compensating factor, distributing loads in such a way that the bearing stresses do not increase, despite temperature changes.

In many systems, disturbances induce two opposing effects that interact to restore stability. This dynamic interplay of compensating actions maintains system balance even in the presence of fluctuating inputs or environmental conditions.

A prominent example of this principle is *regenerative braking*, which increases the level of balancing by capturing *kinetic energy* as a compensating factor. In conventional braking, this energy is lost as heat. In contrast, regenerative braking converts kinetic energy during deceleration into electrical energy, which is stored in batteries or capacitors for future use. This process not only improves energy efficiency by reducing waste but also decreases dependence on mechanical brakes, thereby minimizing heat buildup and wear. The result is enhanced component durability and more consistent system performance under dynamic loading conditions.

Similarly, energy storage systems such as batteries, capacitors, or hydraulic accumulators serve as compensating elements by absorbing and releasing energy during fluctuations, thus levelling out demand and supply mismatches.

Damping in vehicle shock absorbers offers another example. Here, energy is dissipated to counteract oscillations and improve stability, which is a passive compensating mechanism that smooths system response.

In *electronic control systems*, introducing a *controlled time delay* can stabilize dynamic responses. In this case, the compensating factor is the deliberate phase or time shift, which prevents overcorrection and helps maintain steady system behaviour.

11.12 INCREASING THE LEVEL OF BALANCING BY SELECTING COMPONENTS FROM THE SAME VARIETY

Increasing the level of balancing can also be achieved if the variability of critical parameters is reduced.

For an assembly built on components logically arranged in series, the life of the assembly is limited to the shortest life of a component in the assembly. A significant variation in the lives of the components arranged in series shortens the life of the assembly. Thus, variation in the fatigue resistance of components with identical functions, sharing a common load and logically arranged in series, means that the assembly will fail significantly before the expected life span in the absence of such variation. Similarly, the variation in the properties of seals could cause a seal assembly to leak long before such leakage appears in an assembly with small variation of properties.

Reliability-critical parameters vary, and this variability should be controlled to increase the level of balancing. Broadly, the following techniques for controlling variability can be distinguished: (i) selecting components of the same variety; (ii) control of material, structure and properties during manufacturing; (iii) control of geometrical parameters during manufacturing; (iv) control of load magnitudes; (v) control of operating environment and (vi) robust design.

Selecting components from the same variety is an important technique for reducing the negative impact of the variability of properties. Often, deep uncertainty is present about the fractions of components from different varieties. Consequently, an important question is how to maximize the probability of selecting components from the same variety.

Most commonly, the available components are of two critical varieties: *reliable components* and *faulty components*. Usually, the percentage of reliable components in the separate batches (or suppliers) is unknown. In this case, it can be shown that the probability of purchasing all components from the variety 'reliable components' is maximized if all components are sourced from the same, randomly selected batch (supplier).

This can be demonstrated with an example featuring the selection of two reliable components from n large batches (e.g., containing the same type of bearings, seals, electronic components, etc.), which are subsequently built into a system. For the system to work, both components selected must be from the variety 'reliable components'.

Suppose that the fractions of reliable components characterizing the n large batches are $r_1, r_2, ..., r_n$, correspondingly. These are unknown quantities. If each component is purchased from a randomly selected batch with probability (1/n), the probability that both components will be of the variety 'reliable components' is given by $p_1 = \frac{1}{n}\sum_{i=1}^{n} r_i^2$.

If the two components are purchased from two different, randomly selected batches, the probability that both components will be reliable is given by

$$p_2 = \frac{1}{[n(n-1)/2]}\sum_{i<j} r_i r_j = \frac{2}{n(n-1)}\sum_{i<j} r_i r_j$$

It can be shown that $p_1 \geq p_2$:

$$\frac{1}{n}\sum_{i=1}^{n} r_i^2 \geq \frac{2}{n(n-1)}\sum_{i<j} r_i r_j \tag{11.1}$$

Inequality (11.1) is a special case of the classical Muirhead's inequality (Hardy et al., 1999). Its proof can be found in (Todinov, 2019d)

For three batches ($n = 3$), for example, inequality (11.1) becomes

$$(1/3)r_1^2 + (1/3)r_2^2 + (1/3)r_3^2 \geq (1/3)r_1 r_2 + (1/3)r_2 r_3 + (1/3)r_3 r_1 \tag{11.2}$$

The left and right parts of inequality (11.2) evaluate to different values and this difference can be significant. Thus, for $r_1 = 0.9$, $r_2 = 0.2$ and $r_3 = 0.65$, the probability of selecting two reliable components from a single, randomly selected batch is

$$p_1 = (1/3)r_1^2 + (1/3)r_2^2 + (1/3)r_3^2 = (1/3)[0.9^2 + 0.2^2 + 0.65^2] = 0.424$$

while the probability of selecting two reliable components from two randomly selected batches is

$$p_2 = (1/3)r_1 r_2 + (1/3)r_2 r_3 + (1/3)r_3 r_1 = (1/3)[0.9 \times 0.2 + 0.2 \times 0.65 + 0.65 \times 0.9] = 0.298$$

and $p_1 > p_2$.

It must be pointed out that selecting two components from the same batch will also maximize the probability that both selected products will be faulty. This, however, does not mean that, it is more beneficial to select the components from two different batches.

Although selecting from two different batches decreases the probability that both selected components will be faulty, for a faulty system to be present it is not necessary for both selected components to be faulty. Selecting a single faulty component is sufficient. Selecting components from two different batches decreases the probability of having a reliable system (two reliable components). This conclusion remains unchanged if the fractions of reliable components in the previous example are, in fact, fractions of faulty components.

Indeed, if we consider $r_1 = 0.9$, $r_2 = 0.2$ and $r_3 = 0.65$ to be the fractions of faulty components in the batches, the probability of selecting two reliable components from a random batch is

$$p_1 = (1/3)(1-r_1)^2 + (1/3)(1-r_2)^2 + (1/3)(1-r_3)^2 = (1/3)[0.1^2 + 0.8^2 + 0.35^2] = 0.257$$

while the probability of selecting two reliable components from two randomly selected batches is

$$\begin{aligned} p_2 &= (1/3)(1-r_1)(1-r_2) + (1/3)(1-r_2)(1-r_3) + (1/3)(1-r_3)(1-r_1) \\ &= (1/3)[0.1\times 0.8 + 0.8\times 0.35 + 0.35\times 0.1] = 0.132 \end{aligned}$$

Again, $p_1 > p_2$.

No matter what r_i denote (fraction of reliable components or fraction of faulty components), the probability that both components will be reliable is always maximized by selecting both components from the same randomly selected batch. With this, the probability of safe operation is also maximized.

By using the Muirhead's inequality, the result related to selecting two components from n suppliers (batches) can be generalized to selecting $m > 2$ $(m \le n)$ components from n suppliers (batches).

12 Enhancing Reliability and Reducing Risk through Self-Reinforcement

12.1 SELF-REINFORCEMENT MECHANISMS

This chapter introduces a domain-independent method for enhancing system reliability and reducing risk, termed *the method of self-reinforcement*. A defining feature of this method is that *when internal or external forces acting on a system increase, the system's own response intensifies in opposition to these forces. This heightened response reduces the net force driving the system toward failure, thereby improving its capacity to withstand overload and increasing its overall resilience*. As will be shown, self-reinforcement mechanisms can, in many cases, deliver substantial reliability improvements at no additional cost.

While extensive research exists on reliability enhancement through reinforcement of components and systems, self-reinforcement remains largely unexplored. Yet, there is a critical distinction between the two. In conventional reinforcement, additional elements or treatments are introduced to improve performance under stress, but the response from the reinforcement does not increase with increasing the applied load. In contrast, self-reinforcement is characterized by a system's ability to generate a stronger internal response as external or internal forces increase, effectively making the system more resistant as the challenge grows.

This distinction is crucial. In traditional reinforcement, the compensatory effect is static: increasing the magnitude of a stressor does not enhance the reinforcement effect. For example, shot peening of automotive suspension springs introduces compressive residual stresses that help resist fatigue failure. Similarly, cold expansion techniques used in aviation strengthen fastener holes by inducing compressive surface stresses. While both techniques improve reliability, their effects remain fixed regardless of load intensity. Increased loading does not amplify the compressive stresses nor improve the fatigue resistance further. Therefore, while shot peening and cold expansion are examples of conventional reinforcement, they do not exhibit the dynamic, load-responsive behaviour of self-reinforcement.

As a result, mechanisms that do not exhibit a dynamic, load-responsive behaviour should not be labelled as self-reinforcing. The focus of this chapter is specifically on self-reinforcement mechanisms to improve reliability and reduce risk, not on traditional reinforcement methods.

DOI: 10.1201/9781003679301-12

While isolated examples of self-reinforcing behaviour do appear in the engineering design literature, these have not been systematically associated with a unified method of self-reinforcement. For example, Pahl et al. (2007) mention a self-reinforcing effect involving rotational motion, but they do not extend this observation into a general framework or propose self-reinforcement as a cross-domain reliability strategy.

Self-locking devices, such as self-locking screws, grips, hooks, and self-energizing brakes, have been known for a long time. Recent developments include self-locking grips for anchoring fibre-reinforced tendons (Costache et al., 2016). However, such mechanisms have not been interpreted through the lens of self-reinforcement for reliability improvement.

Another example is self-balancing systems, which can inherently provide self-reinforcing responses. A familiar case is the 'Roly-Poly' toy, which returns to an upright position after being tilted. Inspired by this principle, Zhang et al. (2013) proposed a self-balancing mechanism to enhance lateral stability in electric bicycles. Yet, like other cases, self-balancing has not been explicitly recognized as a self-reinforcement mechanism in the context of risk and reliability management.

This chapter identifies three principal types of self-reinforcement:

- *Self-reinforcement via compensating factors*,
- *Self-reinforcement via feedback loops, and*
- *Self-reinforcement through self-service, self-repair, and self-healing.*

Mechanisms that capture a proportional compensating effect yield robust self-reinforcement outcomes. Four fundamental categories of such mechanisms have been identified:

(i) Mechanisms that transform forces or pressure into a self-reinforcing response;
(ii) Mechanisms that convert motion into a self-reinforcing response;
(iii) Mechanisms that generate self-reinforcement through geometric transformation or strain development;
(iv) Mechanisms that achieve self-reinforcement through self-alignment.

The classification presented in this chapter (see Figure 12.1) is based on a broad survey of engineering solutions across various domains, all of which contain elements of self-reinforcement aimed at enhancing reliability and reducing risk (Todinov, 2017c). Each example was carefully analysed to verify and where possible, to quantify the effect of self-reinforcement. Additionally, the examples were examined for common patterns and structural invariants. A deliberate level of abstraction was also applied to strip each solution of its domain-specific details, revealing the underlying self-reinforcement principles.

In the next sections, we provide a detailed discussion of these self-reinforcement mechanisms and their implications for design and reliability engineering.

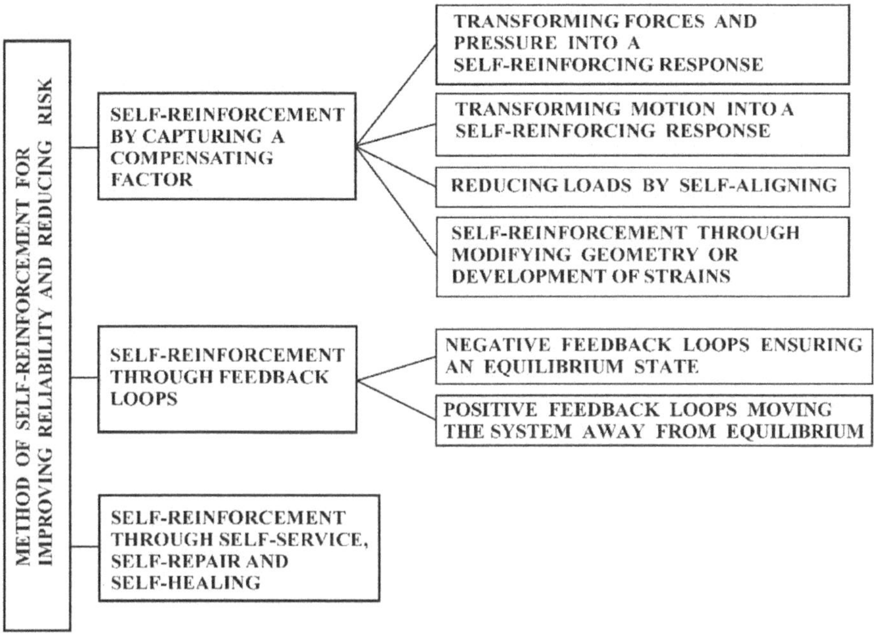

FIGURE 12.1 The method of self-reinforcement for improving reliability and reducing risk.

12.2 SELF-REINFORCEMENT RELYING ON A COMPENSATORY FACTOR

This type of self-reinforcement occurs when a third factor is harnessed to produce a compensating response that strengthens the system. This response counteracts the driving forces that contribute to failure.

12.2.1 Transforming Forces and Pressure into a Self-Reinforcing Response

12.2.1.1 Capturing a Self-Reinforcing Response from Friction Forces

Friction forces can be harnessed as self-reinforcing counterforces. Common examples include self-locking screws and wedges. In such devices, the greater the applied load, the greater the induced friction force, which in turn enhances the reliability of the self-locking mechanism.

This principle is also found in other self-locking systems, such as:

(i) self-locking grips in tensile testing machines,
(ii) self-locking plate clamps,
(iii) self-locking hooks,
(iv) self-locking climbing equipment, and
(v) self-locking marine cleats.

Consider the block in Figure 12.2, subjected to an external force P and influenced by friction. As the magnitude of the force P increases, so does its horizontal component P_x, which can be thought of as an undesirable outcome which tends to disturb the block's equilibrium. According to classical statics (Meriam et al., 2016), the body will remain in equilibrium as long as the resultant force lies within the friction cone defined by the angle ρ ($\mu = \tan \rho$, where μ is the static coefficient of friction).

As the external force P increases, the horizontal component P_x pushing the block also increases. Simultaneously, the normal force $N = P_y$ rises, generating a proportional increase in the friction force F. Since $F = \mu N$, the friction force grows with the normal force, acting as a third compensatory factor that increases proportionally with the undesirable horizontal component P_x, but in the opposite direction. This counterforce F offsets the horizontal disturbance, maintaining equilibrium.

Thus, a self-reinforcing effect emerges. As the external load P increases, so does the resistance to horizontal motion. The conceptual mechanism of self-reinforcement by capturing a proportional compensatory response is illustrated in Figure 12.3. Here, an input factor A leads to an undesirable effect B. Through appropriate design, a counteracting factor C is generated, proportional in magnitude to B but opposite in sign. When the compensatory factor C is combined with B, the net undesirable effect B' is significantly reduced.

This self-reinforcement mechanism improves system reliability by reducing the driving forces that lead to failure. It also enhances resilience by increasing the system's capacity to absorb overload shocks and maintain function under stress.

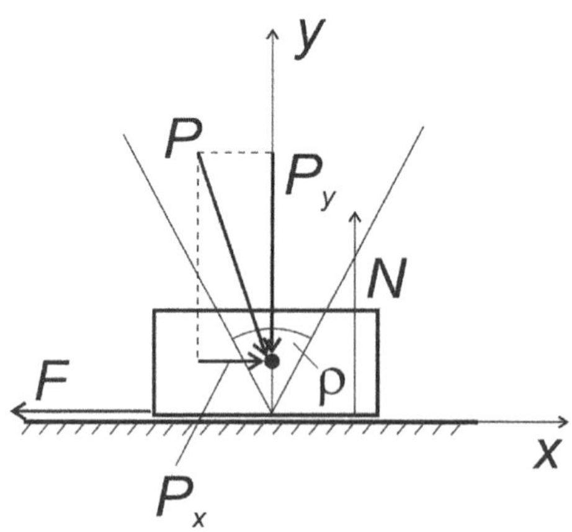

FIGURE 12.2 A cone of friction.

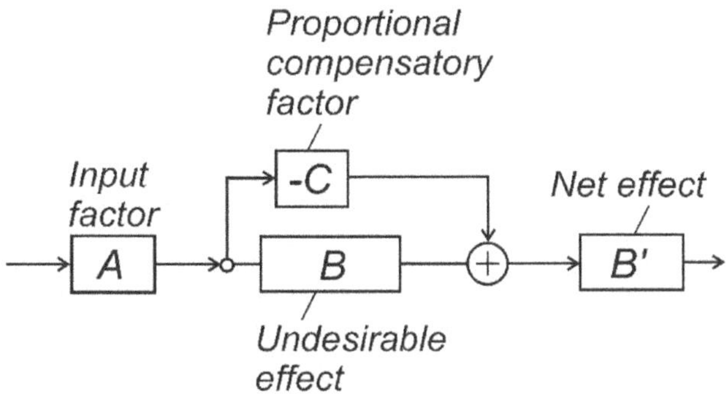

FIGURE 12.3 Self-reinforcement by capturing a compensatory third factor whose magnitude is proportional to the magnitude of the undesirable effect *B*.

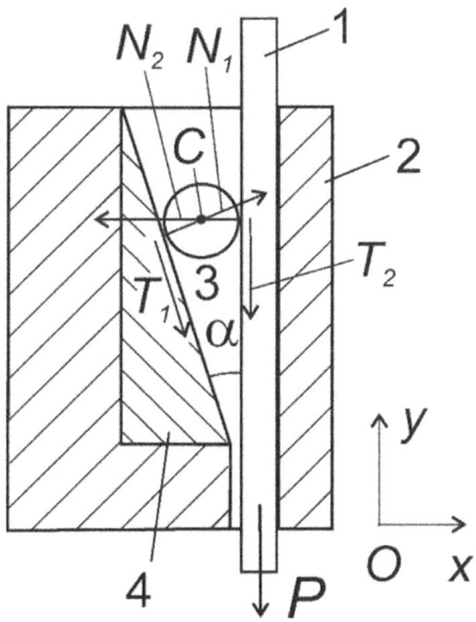

FIGURE 12.4 A self-reinforcing design of a friction grip based on exciting compensating friction forces.

12.2.1.2 Case Study: Transforming Friction Forces into a Self-Reinforcing Proportional Response in the Design of a Friction Grip

Self-reinforcement by capturing a compensatory factor proportional to an undesirable outcome has been implemented in the design shown in Figure 12.4, which illustrates a friction grip mechanism.

The friction grip consists of the following components: strip 1, which is subjected to a downward force *P*; block 2; cylinder 3; and wedge 4. The static friction coefficient between cylinder 3 and wedge 4 is denoted by μ_1, and the friction coefficient between cylinder 3 and strip 1 is denoted by μ_2.

The forces acting on cylinder 3 include:

- A normal force N_1 and friction force $T_1 = \mu_1 N_1$ from wedge 4,
- A normal force N_2 and friction force $T_2 = \mu_2 N_2$ from strip 1.

Under certain conditions analysed in (Todinov, 2017c), the system exhibits *self-reinforcement*, meaning that an increase in the external load *P* proportionally increases the friction forces T_2 and T_1 (represented by factor *C* in Figure 12.3), thus maintaining equilibrium between the strip and the cylinder.

12.2.1.3 Transforming Pressure and Weight into a Self-Reinforcing Response

An example of such a mechanism is presented in Chapter 8, in the design of a pressure container cover (Figure 8.3). The configuration shown in Figure 8.3b channels the force generated by internal pressure into a self-reinforcing action that improves the sealing effectiveness and prevents leakage. As internal pressure increases, the self-reinforcing force, tightening the seal and improving resistance to leakages, also increases.

A similar principle is applied in injection pistons equipped with self-reinforcing seals. Here, the elastic sealing material is pressed more firmly against the cylinder walls as pressure rises. The greater the pressure, the tighter the seal formed between the piston and the cylinder, resulting in improved reliability. In addition, this enhancement in performance is achieved without additional cost.

An example where weight (gravitational force) is channelled or redirected to create a self-reinforcement effect, can be given with the wedge-based friction grips (Figure 12.4). As the weight increases, it enhances the grip strength via increased normal and friction forces which effectively creates a self-reinforcing lock.

Another example where weight is used to create a self-reinforcement effect can be given with the case studies 12.2.2 and 12.2.4.

12.2.2 Case Study: Transforming Moments into a Self-Reinforcing Response

Self-reinforcement through the capture of a compensating, proportionally responsive factor can be illustrated using an example involving a counteracting moment (Figure 12.5). In Figure 12.5a, a wall resists the overturning moment *M* caused by the pressure of a large mass of liquid or granular material (e.g., water, concrete, sand). When an extension is added to the wall's lower end (Figure 12.5b), the distributed weight *G* of the liquid or granular mass generates an opposing moment *M'*, represented as compensatory factor *C* in Figure 12.3, that counteracts the overturning moment (factor *B* in Figure 12.3). This interaction reduces the total overturning moment and, consequently, the risk of overturning the wall. Increasing the mass of the liquid/material increases both the overturning moment and the stabilizing counter-moment, creating a self-strengthening response.

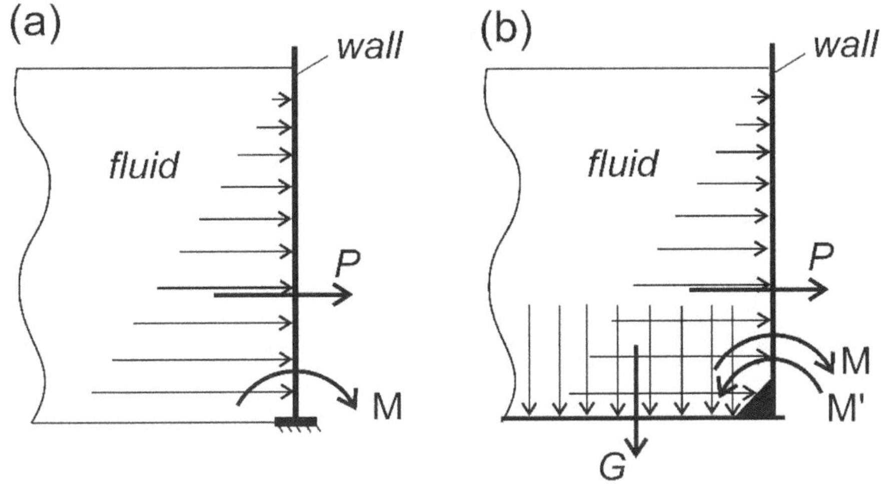

FIGURE 12.5 (a) Overturning moment acting on a wall supporting liquid or granulated substance; (b) reducing the risk of overturning the wall by self-reinforcement.

12.2.3 Self-Reinforcement by Self-Balancing

Unbalanced forces cause premature wear out, fatigue degradation, and failure. As a rule, improving the level of balancing in a system improves the uniformity of the load distribution, reduces the magnitudes of the inertia forces and the loading stresses, and increases the reliability of the system.

The mechanism of self-reinforcement by self-balancing works by using the external or internal forces disturbing the equilibrium of the system to create a self-reinforcing response by increasing the level of balancing of the system. Self-balancing improves the capability of the system to adapt to adversity and recover, which enhances the resilience of the system.

Numerous examples of self-reinforcement through self-balancing have been given in Chapter 11.

12.2.4 Case Study: Self-Reinforcement by Self-Anchoring

Self-reinforcement by self-anchoring is present in cases where the loading forces excite counterforces that oppose the loading forces, which results in a reduced driving force for failure initiation. Self-anchoring bolts and screws are often used as a secure attachment to walls. Increasing the pulling force increases the anchoring effect and the pull-out resistance.

Consider the device in Figure 12.6 designed for lifting steel drums labelled (4), each with mass m. The U-shaped brackets (1) connect the lifting ring (3) to the hooks (2). The device secures itself by hooking over the end lips of the drum. From the equilibrium condition of the forces: $2 \times 2T \times \cos\beta = mg$, the tension force T acting in each rod of the U-shaped bracket can be determined: $T = \dfrac{mg}{4\cos\beta}$. The retaining force

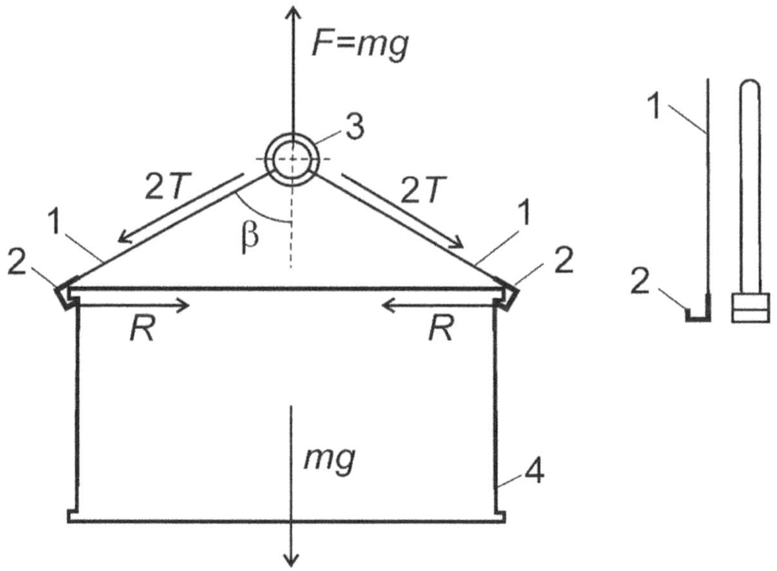

FIGURE 12.6 Hoisting a steel drum by self-reinforcement achieved through self-anchoring.

R with which the hook presses against the drum (as shown in Figure 12.6) can be approximated by $R = 2T \times \sin\beta = \frac{mg}{2}\tan\beta$. As the weight of the drum mg increases, the retaining force R also increases indicating the presence of self-reinforcement.

Increasing the angle β towards 90° results in a significant increase of $\tan\beta$ and the retaining force R. In both cases, reliability is improved through self-anchoring, and this enhancement is obtained at no additional cost.

12.2.5 Transforming Motion into a Self-Reinforcing Response

A classic example is the transformation of motion of a racing car into downforce, particularly through the rear wing of the car (Figure 12.7).

The rear wing functions like an inverted airplane wing, using aerodynamic principles to generate downforce that presses the car onto the track. Its shape causes air to move slower over the top surface than beneath it. According to Bernoulli's principle, faster-moving air beneath the wing creates lower pressure, while slower-moving air over the top surface creates higher pressure. This pressure differential generates a net downward force which increases tire grip. Enhanced grip enables the car to corner at higher speeds, brake more effectively, and accelerate out of turns with improved stability and traction.

This is an example of self-reinforcement obtained by capturing and using motion to provide a compensating, self-strengthening response. The higher the speed, the higher the magnitude of the downforce and the larger is the magnitude of the self-reinforcing response.

Another example of transforming motion into a self-reinforcing response is seen in boats equipped with hydrofoils. As the boat gains speed, water pressure on the

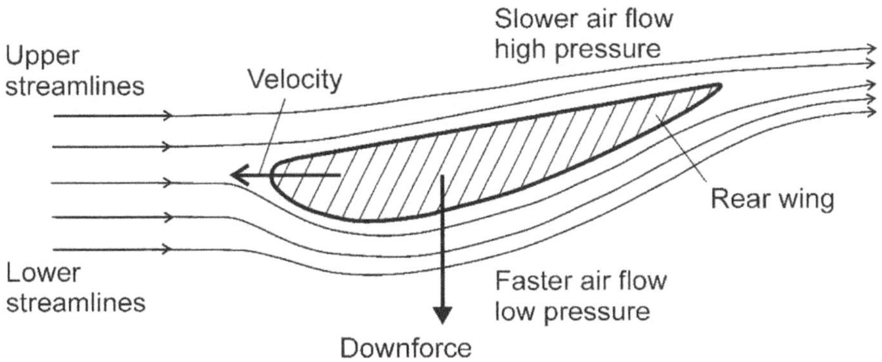

FIGURE 12.7 A rear wing of a racing car transforming motion into a downforce.

hydrofoil generates lift raising the hydrofoil, and consequently, the boat's hull out of the water. This significantly reduces fluid friction, allowing the boat to move faster.

12.2.6 Self-Reinforcement by Self-Alignment

Self-reinforcement through self-alignment operates by harnessing external forces to adjust the position or orientation of an object, thus reducing the resultant force acting on it. This mechanism enhances the system's ability to adapt to adverse operational conditions, improving its overall resilience.

A classic example of self-alignment is the self-aligning ball bearing, which features outer rings with spherical raceways (Bergman et al., 2009). This design accommodates inevitable shaft misalignments. Without the benefit of self-alignment, such misalignments would significantly reduce the bearing's service life.

12.2.6.1 Case Study: Self-Reinforcement by Self-Alignment of a Rectangular Panel under Wind Pressure

Self-reinforcement can also be achieved when the loading force induces self-alignment, thereby reducing its own magnitude. A similar application is discussed in Orloff (2017).

Consider the common case in which the surface of a rectangular panel, with sides L and H, is positioned perpendicular to the wind direction (Figure 12.8a). In this case, the dynamic pressure of the wind $(1/2)\rho v^2$ is transformed into a force $F_w = C_d(1/2)\rho v^2 A_p$ acting on the panel, where C_d is the drag coefficient characterizing the panel, ρ is the mass density of air, v is the air velocity, and A_p is the projected surface area of the panel perpendicular to the velocity of the wind. The wind force F_w can reach a large magnitude because it increases with the square of the air velocity. Thus, wind with velocity v, perpendicular to the panel, whose mass centre G is at a height h above the ground creates the force $F_w = (1/2)\times C_d \times \rho \times v^2 \times A_p$ and the moment

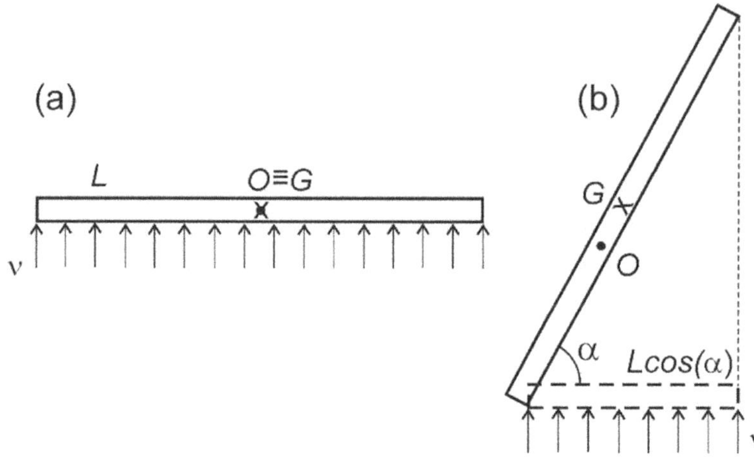

FIGURE 12.8 Self-reinforcement by self-alignment (reducing the stresses at the base of a rectangular panel).

$$M_w = F_w h = (1/2) \times C_d \times \rho \times v^2 \times A_p \times h \qquad (11.8)$$

This may result in a bending moment of considerable magnitude, leading to significant bending stresses at the base of the support. However, if the panel is designed to rotate about an axis (point O in Figure 12.8b) that is offset from the axis passing through the mass centre G, the wind pressure will cause the panel to rotate until the moment created by the wind is counterbalanced by the resisting moment.

Suppose that this counterbalancing occurs at an angle α. In this case, the projected area is $A_p = L \times H \times \cos\alpha$, which results in a bending moment M_w with a significantly smaller magnitude.

Increasing the velocity of the wind produces a self-reinforcing response. Indeed, increasing the air velocity v increases the magnitude of the angle α at which the resisting moment counterbalances the moment rotating the panel. Increasing the magnitude of the angle α reduces the projected area $A_p = L \times H \times \cos\alpha$ which results in a corresponding reduction of the bending moment M_w at the base of the support.

In other words, increasing wind velocity increases the system's response, enabling the system to adapt more effectively to the intensified conditions. This dynamic response enhances the assembly's ability to absorb overloads and operate under adverse conditions. The reliability enhancement is achieved without additional cost.

The underlying principle of a self-reinforcing design can even be applied to everyday objects, such as umbrellas. A well-designed umbrella incorporates flexibility, allowing it to partially collapse and reduce its surface area when facing strong winds. As wind strength increases, the exposed area decreases, which reduces the force exerted on the umbrella.

12.2.7 Self-Reinforcement through Modified Geometry and Strains

Modifying the geometry of loaded components, in some cases creates self-reinforcing effect and leads to a more uniform load distribution, which reduces stresses and enhances the components' load-bearing capacity. For example, specially designed nuts with optimized geometry promote a more even distribution of load along the threads, resulting in higher load-carrying capacity (Matthews, 1998). In fact, as the applied load increases, the load distribution along the thread becomes increasingly more uniform. Similarly, altering a component's shape to a more symmetric shape can induce counterbalancing forces, which contribute to improved structural performance and load-bearing capacity (Matthews, 1998).

In materials, a self-reinforcing effect is observed during plastic deformation, known as strain hardening (Ashby and Jones, 2002). As the material undergoes increasing plastic strain, its resistance to further deformation rises.

Another example of self-reinforcement can be found in Hadfield steels, where deformation-induced strengthening improves resistance to abrasive wear under impact loading (Lindroos et al., 2015). A similar mechanism operates in ceramics, such as cubic zirconia containing dispersed tetragonal zirconia particles. As described by Kelly and Francis-Rose (2002), when a crack propagates in the ceramic, the increased crack-tip opening stress intensifies the local stress field. This triggers a phase transformation of tetragonal zirconia into monoclinic zirconia, which has a larger specific volume. The associated transformation strain generates compressive forces at the crack tip, opposing further crack growth. This is an example of self-reinforcement driven by phase transformation strains.

12.3 SELF-REINFORCEMENT BY FEEDBACK LOOPS

12.3.1 Self-Reinforcement by Creating Negative Feedback Loops

This mechanism employs negative feedback control to achieve self-reinforcement and enhance reliability, as described later. The feedback loop operates as a circular chain of causes and effects, with the negative feedback working to stabilize the system around a defined goal. By continuously sampling the system's output, the feedback loop detects deviations from the target value and adjusts the input accordingly, to reduce these deviations.

If the output exceeds the desired value, the feedback mechanism modifies the input to decrease the output toward the desired value (Figure 12.9a). Conversely, if the output falls below the desired value, the input is adjusted to raise the output (Figure 12.9b). In this way, the system dynamically minimizes deviations and maintains stability around the target output. The greater the deviation from the target output, the stronger the corrective input, demonstrating the self-regulating nature of negative feedback.

Negative feedback loops are widely used in control systems to regulate parameters such as position, temperature, pressure, humidity, concentration, and rotational speed.

A well-known example of a negative feedback loop is the bimetal strip used to provide heat protection of an iron. When the iron is plugged in, the heating element warms up. The bimetal strip is positioned near the heating element, so it heats up too. As the temperature rises, the bimetal strip bends due to uneven expansion. At a preset temperature, it physically breaks the electrical contact, turning off the heating element. Without power, the iron cools down and the strip straightens back out as it cools. Once the strip returns to its original shape, it closes the circuit again, and the iron starts reheating.

Another well-known example of a negative feedback loop is the rumble strips on motorways which help prevent lane departure accidents caused by driver drowsiness, distraction, or inattention. When the vehicle's tyres cross these strips, they produce a loud noise and a strong vibrating sensation, alerting the driver and prompting corrective steering before the vehicle drifts off the road.

Typically installed along lane edges, shoulders, or centrelines, rumble strips function as a passive safety system, particularly valuable at night or in low-visibility conditions, when road markings may be difficult to see. They are relatively inexpensive to install and have been statistically proven to significantly reduce fatal and injury-related crashes on motorways.

Another similar example from road safety is the use of radar speed signs: they measure vehicle speed and display it in real time, providing immediate feedback that helps drivers adjust their speed to comply with legal limits.

A common example of a negative feedback loop is found in the coupling between an asynchronous motor and a driven machine (Figure 12.10). In the stable region of the angular speed (around ω_{st}), if the rotor's angular velocity increases beyond the stable operating point ω_{st}, the driving torque falls below the resisting torque M_{res}, causing the rotor speed to decrease. Conversely, if the rotor's angular velocity drops below ω_{st}, the motor's driving torque exceeds the resisting torque M_{res}, leading to an increase in the rotating speed. This self-correcting dynamics exemplifies the stabilizing effect of negative feedback in electromechanical systems.

In typical negative feedback control systems, the output is continuously measured and fed back to an error detector at the input. A controller then adjusts the system or process parameters to minimize deviations from the desired output, driving the error as close to zero as possible. The widespread availability of low-cost electronic devices and sensors has significantly increased the flexibility of modern control systems, enabling much finer regulation compared to purely mechanical systems.

Negative feedback loops enhance system performance in several ways: they reduce sensitivity to fluctuations in the input signal, improve adaptability to adverse operating conditions, and strengthen the system's ability to recover when equilibrium is disrupted. Together, these effects contribute to greater system resilience.

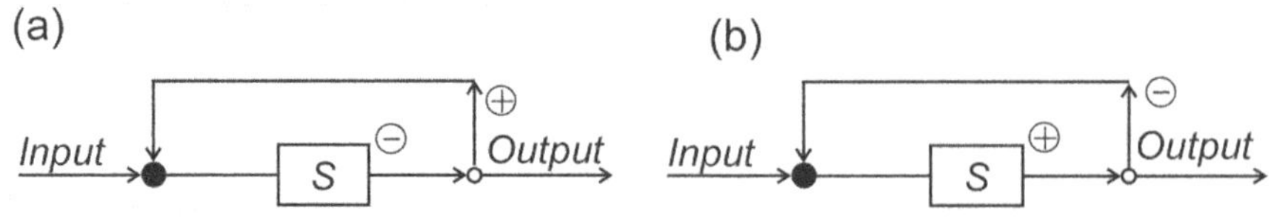

FIGURE 12.9 Negative feedback loop.

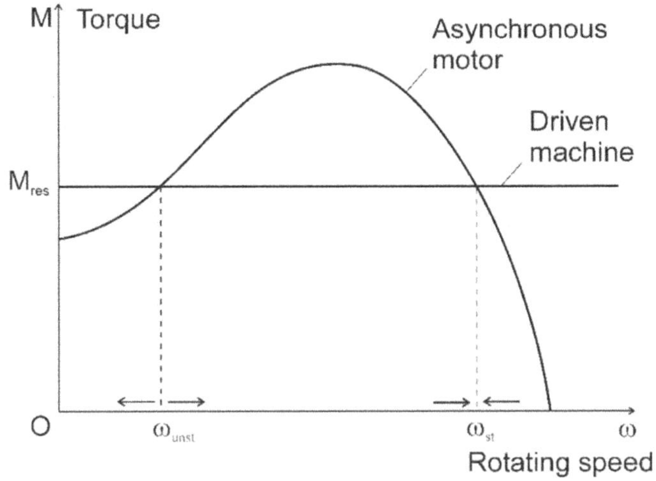

FIGURE 12.10 Output characteristics of an asynchronous motor and a driven machine.

Broadly, negative feedback control can be understood as the interpretation and use of information about a system's state to stabilize its output. Sensor measurements combined with the corresponding corrective responses constitute a typical negative feedback loop.

12.3.2 Self-Reinforcement by Creating Positive Feedback Loops

While the application of negative feedback control loops is well understood and extensively documented, the use of positive feedback loops to enhance reliability and reduce risk remains largely underexplored.

Like negative feedback, positive feedback also operates as a circular chain of causes and effects. However, its mechanism differs fundamentally: in a positive feedback loop, the system responds to deviations by amplifying them rather than correcting them (Figure 12.11a,b). If the output begins to increase, the loop reinforces that trend (Figure 12.11a), leading to a further increase. Conversely, if the output decreases, the loop drives it even lower (Figure 12.11b). In this way, positive feedback mechanisms push the system away from equilibrium rather than stabilizing it around the equilibrium.

Feedback loops are typically represented using causal loop diagrams (Figure 12.12). The arrows in these diagrams indicate the direction of causality, not merely a sequence of events. For instance, a positive link from A to B (Figure 12.12a) implies that 'an increase in *A* leads to an increase in *B*', while a negative link from *B* to *C* implies that 'an increase in *B* leads to a decrease in *C*'.

A positive feedback loop does not necessarily consist of a direct connection from output back to input. It may involve a longer chain of causal relationships, culminating in an effect that reinforces the initial input. For example, in the loop shown in Figure 12.12a, an increase in input factor *A* causes an increase in factor *B*, which

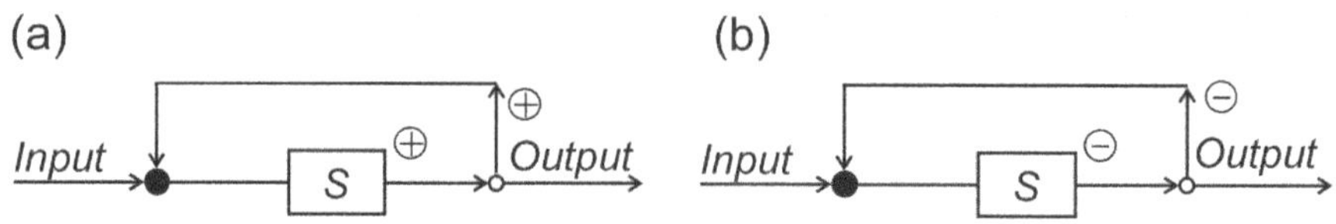

FIGURE 12.11 Positive feedback loop.

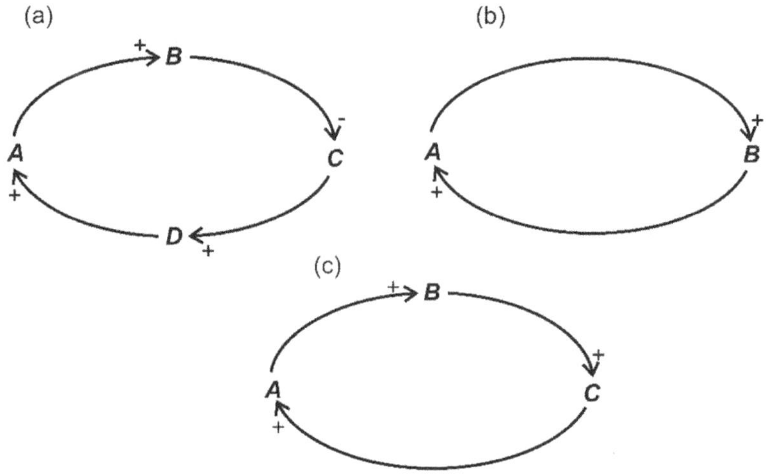

FIGURE 12.12 Positive feedback loop diagrams.

then leads to a decrease in factor *C*. This decrease, in turn, causes an increase in factor *D* which feeds back to a further increase of factor *A*. The net result is that an initial increase in *A* ultimately causes a further increase in *A*, demonstrating the self-reinforcing nature of positive feedback.

When a system has an initial base level, even a slight deviation from this baseline can trigger further deviation in the same direction through a positive, self-reinforcing feedback loop. In this sense, the positive feedback mechanism amplifies the gap between the system's current state and its base state. A positive deviation initiates further positive deviation, resulting in a process of growth. Conversely, a negative deviation leads to further decline, resulting in a process of decay.

By their nature, positive feedback loops tend to destabilize a system's output, driving it away from equilibrium. They are often cited as underlying mechanisms in phenomena such as prolonged droughts, global warming, financial crises, and bank runs. Because of these associations, positive feedback loops are commonly perceived as inherently destructive and as precursors to system collapse.

However, positive feedback loops can also serve as powerful engines for beneficial growth and improvement. When properly harnessed, they can contribute to enhanced system reliability and effective risk reduction.

There are two principal strategies for leveraging positive feedback loops to manage risk:

- Suppressing or disrupting self-reinforcing feedback loops with negative impact.
- Encouraging and using self-reinforcing feedback loops that yield beneficial outcomes.

12.3.2.1 Reducing Risk by Eliminating or Inhibiting Self-Reinforcing Positive Feedback Loops with Negative Impact

Significant risk reduction can be achieved by identifying and eliminating self-reinforcing positive feedback loops that destabilize a system. A classic example of such a loop with negative impact is an exothermic chemical reaction. The heat generated during the reaction (cause *A*), if not effectively removed, accelerates the reaction rate (effect *B*), which in turn produces even more heat (cause *A*), and so on (Figure 12.12b). This escalating cycle can lead to thermal runaway and, ultimately, to a chemical explosion.

Positive feedback loops of this nature are present in lithium-ion batteries, widely used as power sources in electronic devices and electric vehicles. A critical parameter for the lithium-ion batteries is the thermal runaway onset temperature, which is the point at which exothermic side reactions begin and trigger uncontrolled temperature increase. This threshold can be reached due to events such as internal short circuits caused by physical damage, overheating, or overcharging (Feng et al., 2017). Eliminating the potential for such runaway reactions by redesigning chemical processes, can significantly reduce risk by removing destabilizing feedback mechanisms.

Several strategies exist to interrupt or prevent this dangerous feedback loop in lithium-ion batteries. For instance, temperature-sensitive additives can be used to shut off current flow at high temperatures, effectively introducing a negative feedback loop that stabilizes the system (Feng et al., 2017). Similarly, effective thermal management, such as active cooling to maintain cell temperatures below critical levels, or the use of thermally stable materials can inhibit feedback escalation.

An advanced battery management system also plays a critical role. A well-designed battery management system ensures continuous monitoring and regulation of temperature, which reduces significantly the likelihood of thermal runaway.

It is important to note that under normal conditions, a negative feedback loop helps maintain thermal stability. As the cell temperature rises, the resulting thermal gradient enhances heat dissipation, preventing further temperature increases. The system remains in stable equilibrium. However, once the temperature exceeds a critical threshold, the heat generation outpaces heat dissipation, and the positive feedback loop dominates, leading to thermal runaway.

Interconnected networks often exhibit self-reinforcing positive feedback loops that can have a profoundly negative impact. In coupled power and telecommunication networks, such a loop can emerge because the telecommunication system is powered by the electrical grid, while the power grid is controlled and monitored through the telecommunication infrastructure. A failure of several nodes in the power grid due to random overload-induced faults (Event A in Figure 12.12b) can lead to the failure of dependent nodes in the telecommunication network (Event B). In turn, these telecommunication failures impair control of the power grid, leading to further failures in the power network (Event A). This destructive cycle may continue until both networks collapse.

This example illustrates how positive feedback loops between interdependent systems can create systemic vulnerabilities. One way to mitigate this risk is to temporarily decouple the networks during a crisis, break the feedback loop and restore stability. As system interdependency increases globally, the importance of identifying and eliminating such destabilizing feedback loops becomes even more critical.

Positive feedback loops are not limited to large-scale networks. For instance, in electronic circuits, unwanted positive feedback due to capacitive or inductive coupling can cause spontaneous oscillations, compromising amplifier stability. Debugging such circuits often involves removing or neutralizing these loops to enhance performance and reliability.

In structural engineering, overloaded structures can also experience positive feedback loops. When certain load-bearing elements fail due to excessive stress, the remaining elements must carry a higher load which further increases the loading stress. This accelerates the deterioration of the load-bearing elements, leading to further failures which further increase the load on the remaining components. This cycle continues until the entire structure collapses.

A compelling analogue exists in organizational systems. Suppose an overworked expert leaves an organization due to job dissatisfaction (Factor A in Figure 12.12b). If no replacement is hired due to poor management strategy, the responsibilities of the leaving expert are redistributed among remaining staff which increases their workload. As a consequence, the increased workload (Factor *B)* leads to further job dissatisfaction (Factor *A*) among remaining staff, prompting more experts to leave. If unchecked, this self-reinforcing loop can deplete the organization of skilled personnel, severely degrading its performance.

Human behaviour also demonstrates harmful positive feedback loops, particularly between belief and choice. For example, widespread belief in a flawed model initiates its widespread adoption. This choice in turn, reinforces belief in the model despite clear experimental data contradicting its predictions. This feedback between belief and choice entrenches a false modelling paradigm, resistant to contradicting evidence from experiments, simulations, and theoretical critiques.

In the economic domain, political instability can initiate a destructive positive feedback loop. A crisis may trigger capital flight, reduced investment and increased emigration, which deepens poverty. The resulting economic decline further reduces investment and increases emigration, perpetuating the cycle of instability and underdevelopment.

12.3.2.2 Self-Reinforcement by Creating Positive Feedback Loops with Positive Impact

Positive feedback loops with beneficial impact are self-reinforcing loops that promote and sustain growth. A typical example can be given from reliability engineering: an initial investment in reliability improvement (*A*) leads to a superior product

(*B*), which in turn results in increased sales (*C*). The revenue from increased sales is then reinvested in further reliability improvements (*A*). This cycle, illustrated in Figure 12.12c, creates a virtuous loop that drives sustained product enhancement, characterized by increased reliability and sales profit.

To maximize the benefits of such positive feedback loops, accelerators can be employed. In a business context, these loops act as powerful engines of growth. For instance, a company creates value that attracts customers. The resulting profits are reinvested, further increasing value creation and number of customers.

Customer surveys serve as effective accelerators in this loop. Feedback gathered through surveys enables the company to tailor its offerings to meet customer expectations, enhancing service quality, expanding the customer base, and fuelling further growth.

Positive feedback loops also play a vital role in bistable systems, enabling rapid transitions between stable states. If the system input slightly exceeds the equilibrium, the feedback drives the output further in the same direction until the upper stable state is reached. Conversely, if the input falls just below the equilibrium, the system rapidly transitions to the lower stable state. Once the system settles into a stable state, it remains there. This behaviour is useful in electromechanical switches, where positive feedback control shortens the arcing period during switching, thus reducing wear and extending the life of the switch.

Additionally, positive feedback loops are fundamental in electronic circuits that trigger specific actions when an input signal crosses a defined threshold. The Schmitt trigger, for example, employs positive feedback to force the output into one of two stable logic states, ensuring clear and noise-resistant digital signal transitions.

12.3.2.3 Case Study: Positive Feedback Loop Providing Self-Reinforcement by Self-Energizing

Positive feedback can also be effectively utilized to improve system reliability and reduce the risk of failure. A practical example of self-reinforcement through a positive feedback loop can be given with another variety of a self-energizing friction holder including rotating friction pads (Figure 12.13) that support the weight G of a suspended object. In this mechanism, similar to the friction grip from Figure 12.4, the load itself enhances the holding force. The greater the weight, the stronger the friction grip, which demonstrates a beneficial positive feedback loop that enhances the reliability of the grip.

Once the friction pads of the holder touch the sheet, a small normal force N appears and excites a proportional friction force $F_{fr} = \mu N$ where μ is the coefficient of friction between the friction pad and the object. The friction force $F_{fr} = \mu N$ is causing rotation of the links around the pins. However, the geometric constraint causes the normal force N to increase which in turn increases the friction force $F_{fr} = \mu N$ and so on. This process of self-excitation continues until the friction forces $F_{fr} = \mu N$ acting on both sides of the object counterbalance the weight G of the object.

As a result of the positive feedback loop, the friction forces $F_{fr} = \mu N$ quickly reach large magnitudes even though the friction pads were initially energized with a very small load. The result is a reliable support of the weight G of the object.

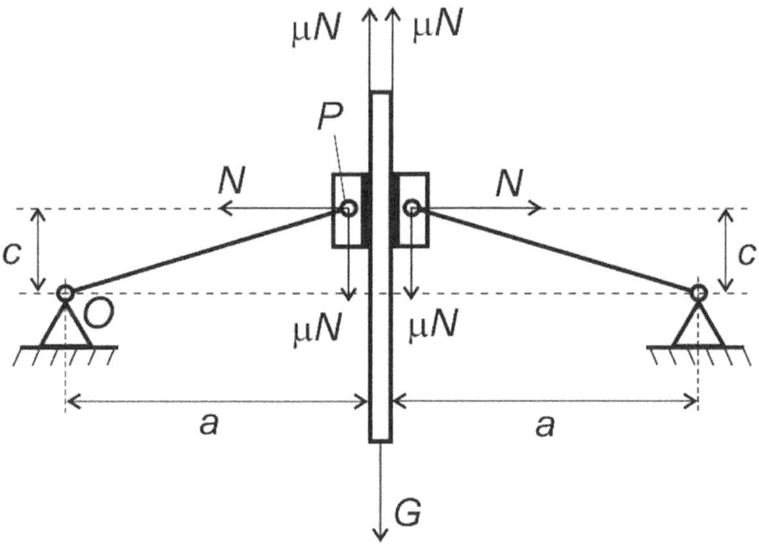

FIGURE 12.13 Self-energizing friction holder including rotating friction pads.

The smallest coefficient of friction μ^* necessary to achieve self-energizing can be determined from the equilibrium condition of the link OP. From the equilibrium condition of the moments with respect to the pin joint O:

$$Nc - \mu Na = 0$$

is obtained, which is equivalent to $c - \mu a = 0$. Self-energizing is present if $\mu \geq c / a$. The smallest coefficient of friction for a self-energizing grip is given by $\mu^* = c / a$.

12.4 SELF-REINFORCEMENT THROUGH SELF-SERVICE, SELF-REPAIR, AND SELF-HEALING

Autonomic computing systems are prime examples of self-servicing technologies. These systems are designed with software components capable of detecting, diagnosing, and repairing faults without human intervention. By doing so, they enhance the reliability and uptime of critical infrastructures such as cloud computing platforms, data centres, and enterprise systems.

A well-known case is the cloud infrastructure auto-healing, where applications automatically detect failures and rectify them without user involvement, ensuring continuous service availability for mission-critical operations.

Another example is self-servicing industrial equipment. Intelligent CNC machines and 3D printers, for instance, often include built-in capabilities for tool wear detection and self-recalibration. These features help maintain high manufacturing quality while minimizing unexpected downtime.

Smart vehicle diagnostics represent yet another domain of self-servicing systems. Modern electric vehicles routinely perform self-diagnostics, recommend maintenance, apply software updates and, in some cases, carry out basic self-repairs. Such features reduce downtime, improve reliability, and lower the potential for human error. Some electric vehicles can even receive software-based performance enhancements remotely.

In robotics, certain systems are equipped with self-repair capabilities. Modular robots, for example, can reconfigure themselves when one or more of their components fail. This self-repairing behaviour is critical for mission resilience in hazardous environments such as planetary exploration or search and rescue operations.

Some self-repairing technologies draw inspiration from biology. An example can be given with synthetic skin used in robotics, which can sense and heal small cuts. These biologically inspired systems pave the way for more adaptive and resilient technologies.

Self-healing materials also play a vital role in aerospace and civil engineering. Certain polymers, for example, are embedded with microcapsules that release healing agents when cracks develop. This innovation helps prevent the propagation of microcracks, significantly extending the lifespan of aircraft components, concrete bridges, and space structures. Some space exploration agencies, in particular, use self-healing composites in spacecraft panels to prevent catastrophic damage from micrometeoroid impacts.

Fuel tanks in some fighter aircraft and modern armoured vehicles are equipped with multi-layered, self-sealing liners. When a bullet or projectile punctures the tank, the liner material is immediately exposed to fuel. Upon contact, the material swells and expands, quickly sealing the breach. This rapid reaction helps prevent severe fuel loss and greatly reduces the risk of fire or explosion.

Lastly, smart grid self-healing networks exemplify self-servicing in electrical infrastructure. These grids automatically detect faults, isolate affected sections, and reroute power as needed. As a result, they greatly enhance grid reliability, reduce outage durations, and minimize the consequences of component failures. Power is often restored autonomously without the need for manual intervention.

13 Tutorial Exercises on Reliability Improvement and Risk Reduction

To improve understanding and mastery of the domain-independent methods presented in the book, this chapter includes a number of tutorial exercises. Each exercise is designed to be solved using one or more of the methods discussed in the book. Detailed solutions to the tutorial exercises are provided in the next chapter.

To extract the maximum benefit from these exercises, each one should be attempted independently before referring to the solutions.

The general approach for developing a reliability improvement solution using domain-independent methods is outlined in Figure 2.2. A key step is to review the methods and techniques covered in the book sequentially and determine which method, technique, or combination thereof could eliminate the identified critical failure mode.

When presenting a solution, it is essential to provide strong, well-reasoned arguments that demonstrate how the solution addresses or mitigates the identified critical failure mode. Additionally, the proposed design must be carefully analysed to ensure it does not introduce new critical failure modes.

If a new critical failure mode is introduced by the proposed solution, its cause must be identified. The next step is to again review the domain-independent methods and techniques in the book, to determine which method or combination of methods could eliminate the new failure mode.

It is also important to ensure that the benefits of the proposed solution justify its implementation costs. Ideally, the solution should improve reliability or reduce risk without significantly increasing cost, mass, energy consumption, or spatial requirements. However, a modest increase in cost may be acceptable if the expected reliability gains significantly outweigh the added expense.

Finally, the proposed solution must not have a negative impact on the environment.

Here is the set of tutorial exercises:

Exercise 13.1

By using the bottom-up approach, discover the failure modes of the clevis joint in Figure 13.1. The clevis is subjected to a constant axial tensile loading force *P*.

 DOI: 10.1201/9781003679301-13

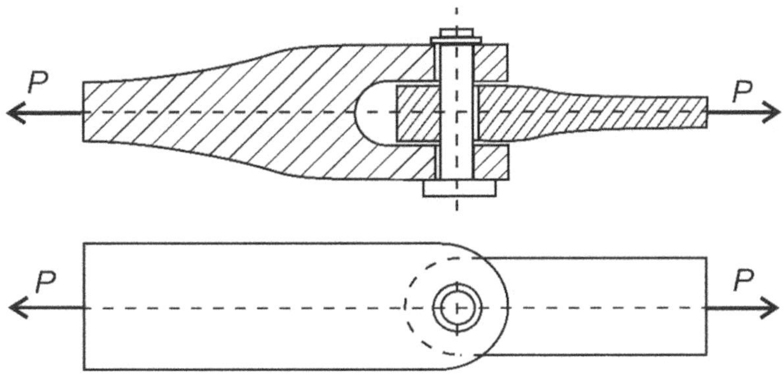

FIGURE 13.1 A clevis joint.

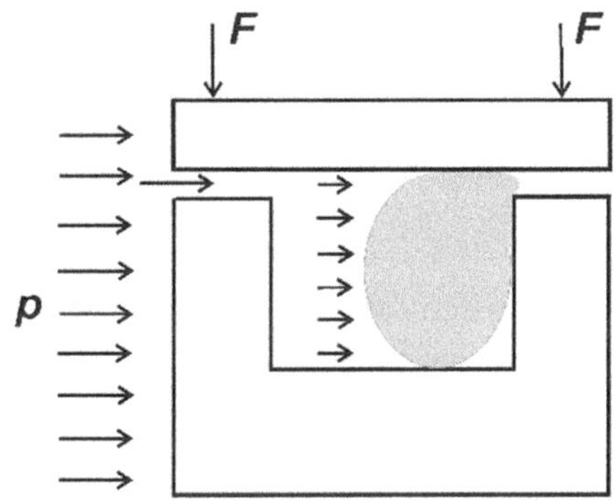

FIGURE 13.2 O-ring seal.

Exercise 13.2

Use the top-down approach to discover the root causes contributing to the system failure mode 'leakage from an O-ring seal' (Figure 13.2). Draw the Ishikawa diagram for the root causes of this failure mode.

Exercise 13.3

Before examining the fracture surface of failed suspension springs using a scanning electron microscope to determine the cause of failure, it is essential to preserve the integrity of the surface. Cleaning the fracture surface with mechanical brushes may damage critical features and distort the microscopic evidence. By using a domain-independent method for improving reliability, propose a cleaning technique that does not damage the fracture surface and distort the microscopic evidence.

Exercise 13.4

Traditional bonded strain gauges are highly sensitive to the quality of surface preparation and adhesive application. As a result, significant measurement variability can

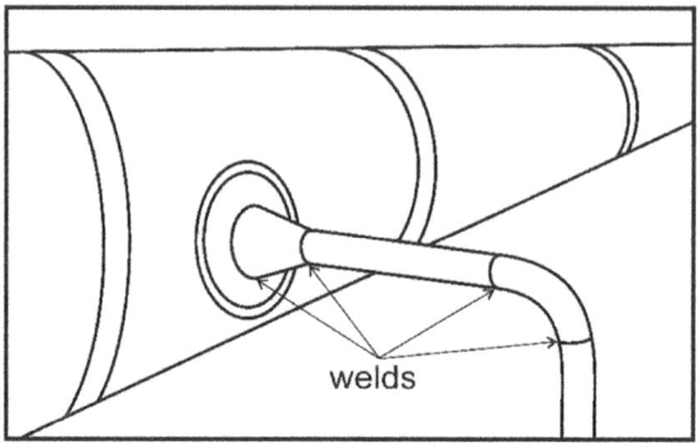

FIGURE 13.3 A system of pipes carrying odourless and colourless gas under pressure.

occur due to inconsistencies in the attachment process. Poor bonding between the gauge and the surface often leads to distorted or unreliable readings. Moreover, the measurement process is slow, as it requires meticulous surface preparation and secure attachment of the gauges. Propose a solution based on a domain-independent method that eliminates these failure modes.

Exercise 13.5

The welds on a system of pipes carrying odourless and colourless gas under pressure (Figure 13.3) can be faulty and cause leaks. By using a domain-independent method, suggest a reliable and inexpensive method of testing the welds for leaks and identifying the faulty welds.

Exercise 13.6

This example is adapted from Starovoytova (2021).

The pressure of the inert gas inside a light bulb is lower than atmospheric pressure (approximately 80% of the atmospheric pressure when the light bulb is cold). Microcracks in the glass envelope allow air to enter the bulb, increasing the internal pressure beyond normal levels. The intrusion of air leads to oxidation of the filament, resulting in premature failure of the light bulb. By using a domain-independent method, suggest an inexpensive method of identifying the faulty light bulbs.

Exercise 13.7

In the initial design (Figure 13.4), both the container (1) and the cylindrical hinges (2) are made of the same material, which has relatively low fatigue resistance. An asymmetric lid (not shown) rotates about the hinges and subjects them to a pulsating force, F. Due to this cyclic loading and the low fatigue resistance of the material, cracks develop at the base of the hinges, eventually leading to failure. By using a domain-independent method covered in the book, improve the reliability of the container.

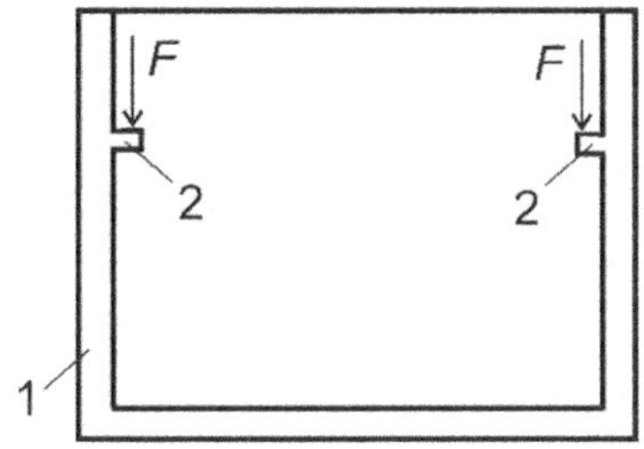

FIGURE 13.4 A container with rotating lid, made of homogeneous material.

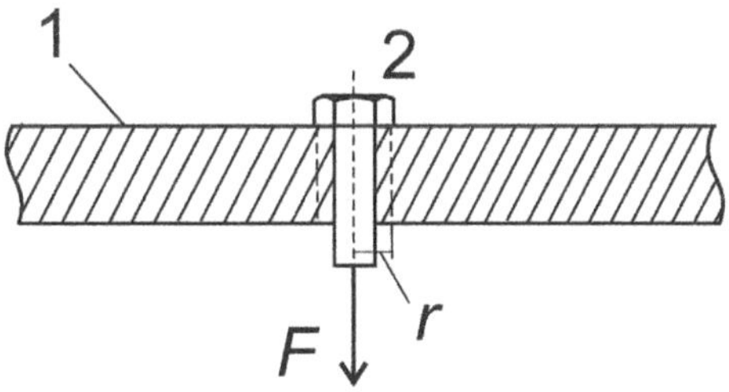

FIGURE 13.5 Improving the reliability of a connection inserted through a low-strength component.

Exercise 13.8

A concentrated force *F* is applied directly on a component (2) (Figure 13.5) inserted into a low-strength material (1) (e.g., a bolt inserted into a ceiling made of thin panels). For a large pulling force *F*, the excited shear stresses along the surface of the cylinder (marked by dashed lines) easily overcome the resistance of the material and cause shear failure. By applying a domain-independent technique, improve the reliability of the connection.

Exercise 13.9

The primary hazard faced by operators of metal-cutting guillotines (also known as metal shears) is the risk of hand amputation or severe lacerations.

In addition, crushing injuries may occur if hands become trapped beneath the clamping bar or during blade descent. This typically happens when an operator attempts to adjust or remove material while the machine is still in motion.

Using a domain-independent method covered in the book, design a safety mechanism that reliably prevents the operator's hands from entering the cutting zone of the guillotine.

Exercise 13.10

The adhesive bond (1) in Figure 13.6 exhibits a good shear strength and distribution of the stresses but unsatisfactory peel resistance. By using an appropriate domain-independent method, improve the reliability of the adhesive joint.

Exercise 13.11

This exercise is adapted from Starovoytova (2021). The pipe-transporting pellets in Figure 13.7 suffers a critical failure mode: Intensive wear in the region of the bend, which leads to fast wearout and failure of the pipe. By using an appropriate domain-independent method, improve the reliability of the pipe.

Exercise 13.12

This exercise is adapted from Erhard (2006).

A plastic sheath must be fastened to a steel plate using screws (Figure 13.8). The identified critical failure mode is damage to the plastic sheath caused by overtightening the screws. Using a domain-independent approach, propose a solution that eliminates this failure mode.

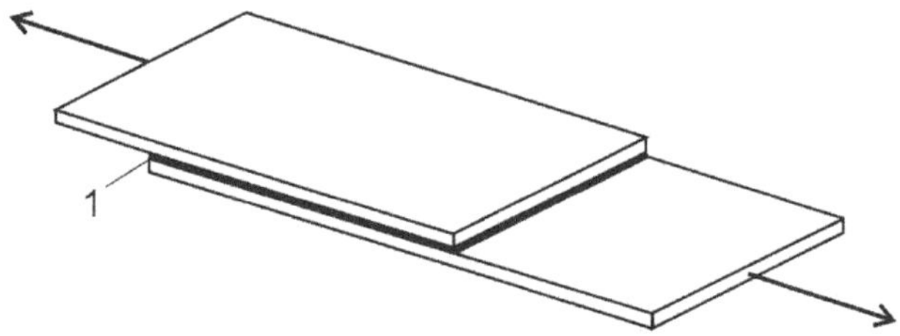

FIGURE 13.6 Adhesive bond.

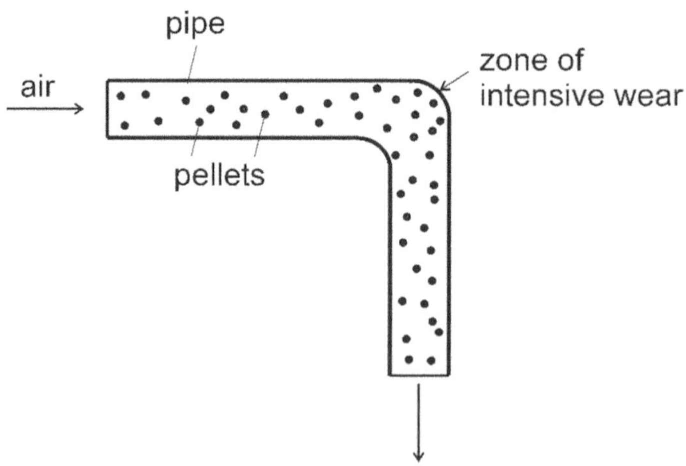

FIGURE 13.7 A pipe with a bend transporting pellets.

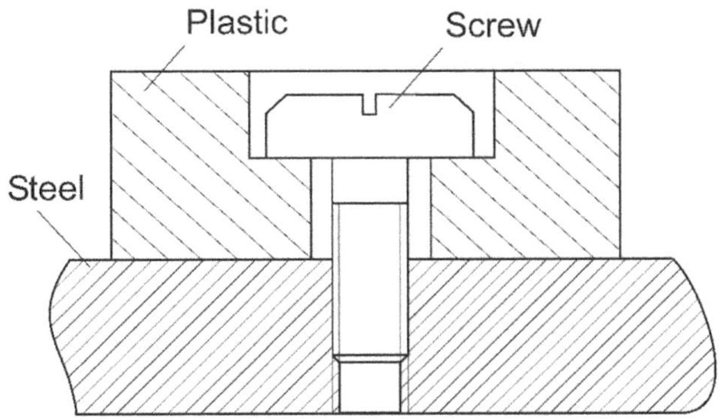

FIGURE 13.8 Plastic sheath fixed to a steel plate.

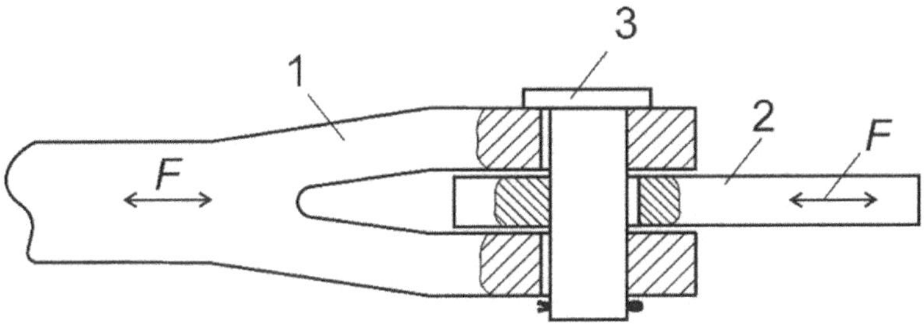

FIGURE 13.9 A clevis joint subjected to a pulsating force *F*.

Exercise 13.13

This exercise is adapted from Rantanen and Domb (2008).

The clevis joint in Figure 13.9 is subjected to a pulsating force *F*. As a result, the parts of the clevis joint (clevis (1), eye (2), and pin (3) are subjected to intensive wear which eventually results in premature failure. By using a domain-independent method, redesign the clevis joint to eliminate the intensive wear.

Exercise 13.14

Suppose that the single light source in Figure 13.10 is used for illuminating a critical manufacturing process. Failure of the light source entails an interruption of the manufacturing process and causes severe delays and lost production. By using a domain-independent method, reduce the vulnerability to failure of the light source.

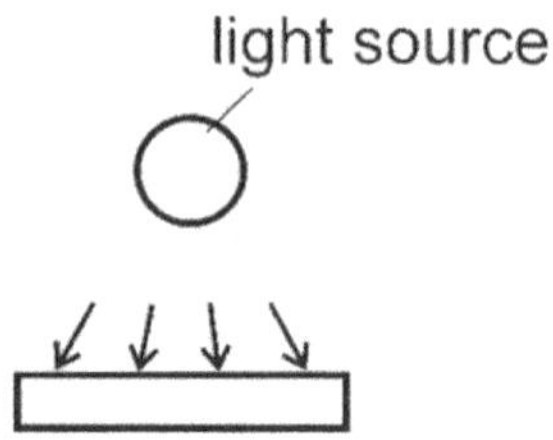

FIGURE 13.10 A light source illuminating a critical manufacturing process.

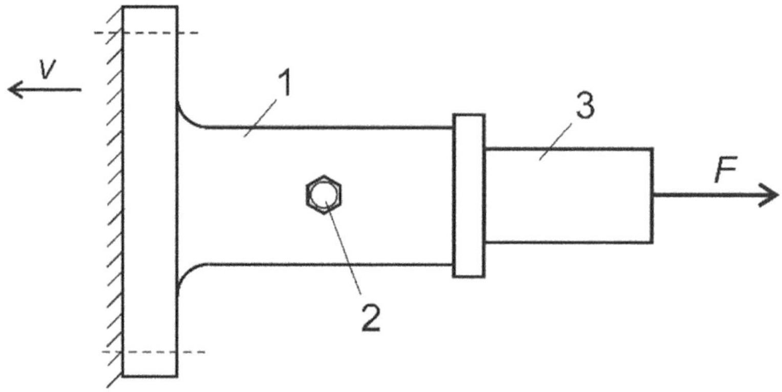

FIGURE 13.11 A trailer hitch assembly connecting a hitch receiver (1) and a trailer hitch (3).

Exercise 13.15

On many towing vehicles, the hitch receiver (1) is connected with a single bolt (2) to the trailer hitch (3) (see Figure 13.11). If this bolt were to fail, the trailer receiver (1) and the trailer hitch (3) will be disconnected, and the trailer will be left on safety chains only. Connection of the towing vehicle and trailer with safety chains only reduces the control over the trailer and is a frequent source of accidents. By using a domain-independent method, improve the safety of this assembly.

Exercise 13.16

In modern commercial or military aircraft, continuous electrical power is essential for flight control, avionics, fuel systems, and life-support systems. A single electrical bus might distribute power to all critical systems. If this single bus fails (e.g., due to a short circuit or generator fault), multiple critical systems could lose power simultaneously. By using a domain-independent method, reduce the vulnerability of the single bus power supply.

Exercise 13.17

Robotic systems used in manufacturing, bomb disposal and exploration, often experience failures in their toolheads, arms, and mobility bases. By using a domain-independent method covered in the book, reduce the downtime caused by these failures, thereby increasing production availability.

Exercise 13.18

A *critical flaw* is a defect that, if present in a brittle column, can initiate an unstable crack that propagates through the column material. In a monolithic brittle column, the presence of a critical flaw typically leads to catastrophic failure. By using a domain-independent method covered in the book, improve the fault tolerance of the column in Figure 13.12 without changing the material of the column.

Exercise 13.19

In systems where torque is transmitted between a rotating shaft and a gear, sprocket, or pulley, retention devices such as keys (Figure 13.13) are commonly used to prevent relative rotation between components. For a given torque, the contact stresses on a single key can be quite high due to the load concentration over a small contact area. Reduce the likelihood of failure of the key assembly and mitigate the risk of over-stress by applying a domain-independent method.

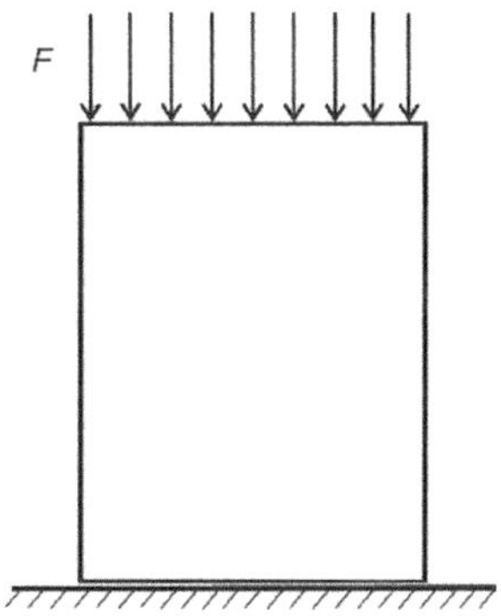

FIGURE 13.12 A monolithic brittle column loaded with a distributed force.

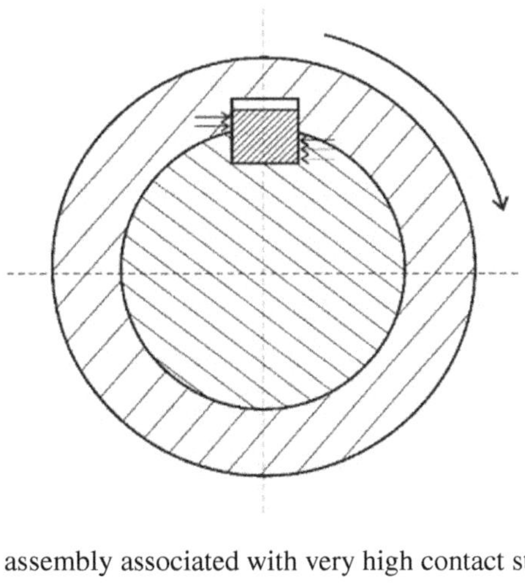

FIGURE 13.13 A key assembly associated with very high contact stresses.

Exercise 13.20

Hoist ropes must be strong, flexible, and lightweight to ensure efficient operation. However, a single rope may fail due to wear, corrosion, overload, or accidental damage. Rope failure can result in a dropped load, causing injury to personnel and damage to equipment. Using a domain-independent method, propose a solution to reduce the likelihood of this failure mode.

Exercise 13.21

This exercise is adapted from Salamatov (1999).

Copper and manganese are essential micronutrients that support enzyme activity and metabolic functions in cattle. To ensure these elements are present in the forage, they must be added to fertilizers or irrigation water in very low, evenly distributed amounts.

Mechanical blending often fails to distribute these elements evenly because micronutrients are added in gram-per-hectare quantities, while bulk fertilizers are in hundreds of kilograms. As a result, some plants may receive an excessive dose, while others get too little, leading to inconsistent nutrient levels in the forage. By using a domain-independent method, suggest a reliable method of dispersing the elements evenly.

Exercise 13.22

This exercise is adapted from (Terninko et al., 1998).

It is essential for a vice to securely clamp workpieces with complex shapes, in order to eliminate the risk of instability during operations (Figure 13.14). Using a domain-independent design approach, how would you develop a vice capable of reliably clamping workpieces of arbitrary shape?

Exercise 13.23

Patio door locking mechanisms are susceptible to failure when the door handle is slammed forcefully. A damaged mechanism typically requires replacement, which

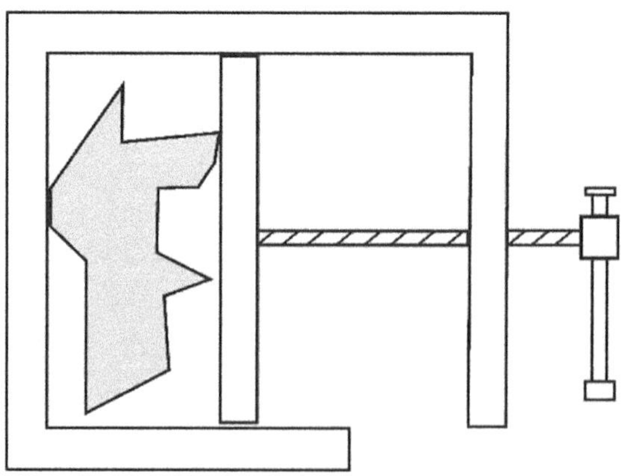

FIGURE 13.14 A vice that cannot clamp securely workpieces with irregular shape.

entails the high cost of both the locking component and locksmith services. By using a domain-independent method, eliminate or significantly reduce the risk of failure of the locking mechanism.

Exercise 13.24

A conveyor belt transports paving slabs through the polishing section of a production line. Over time, the belt experiences significant wear due to the pressure exerted by the polishing heads on the slabs. Propose a domain-independent solution to reduce the repair and maintenance costs associated with belt degradation.

Exercise 13.25

In a software system that provides a specific service, an excessive number of incoming requests can lead to system failure. Propose a solution, based on a domain-independent method, to prevent the system from collapsing under high load.

Exercise 13.26

In an assembly line, the drive mechanism of a robotic arm is susceptible to damage if the resisting torque exceeds a certain threshold. Propose a solution, based on a domain-independent method, to reduce the likelihood of such a failure mode.

Exercise 13.27

Luggage wheels are subject to several well-known failure modes:

- wheel delamination;
- wheel denting;
- wheel wear;
- wheel breakage;
- wheel mount breakage.

These failure modes typically result from shock loading that occurs when luggage is pulled over uneven surfaces. Propose an effective solution, based on a domain-independent method, to prevent or mitigate these failure modes.

Exercise 13.28

Water-filled vessels are susceptible to a critical failure mode: rupture due to the expansion of water when it freezes. Propose a robust solution, grounded in a domain-independent method, to prevent this failure.

Exercise 13.29

The brackets (1) in Figure 13.15 support a component (2) subjected to a varying force F caused by an unbalanced rotating mass. The critical failure mode is fatigue failure of the supporting brackets. Propose a solution, based on a domain-independent method, to extend the fatigue life of the supporting brackets.

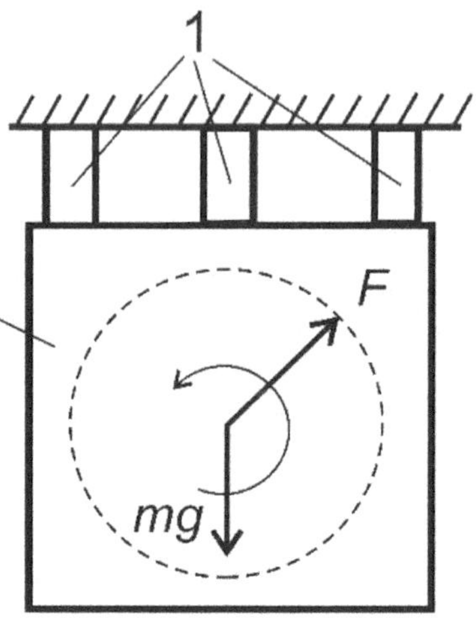

FIGURE 13.15 Brackets supporting a component subjected to a varying force.

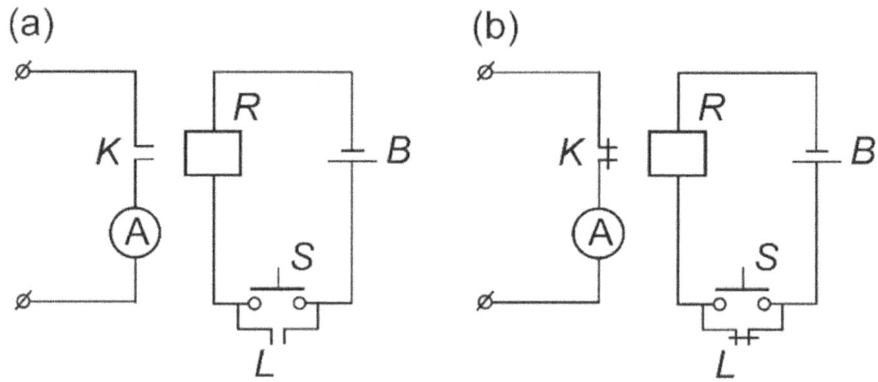

FIGURE 13.16 An alarm system with critical failure modes.

Exercise 13.30

Suggest a reliable method to test the seal on a small container containing liquid. Base your solution on a domain-independent method.

Exercise 13.31

Consider the alarm circuit shown in Figure 13.16a, which is activated by a normally open switch *S*. When the switch *S* is pressed, it energizes the latching relay *R*, causing its normally open contact *K* and latching contact *L* to close. As a result, the alarm remains activated after the switch *S* is released (Figure 13.16b).

However, there is a critical failure mode associated with the design in Figure 13.16. The circuit is vulnerable to open-circuit failures and loss of power from the source B. Specifically, if the control circuit becomes open, the wires connected to switch *S* are severed, or the power source *B* fails (e.g., a dead battery), pressing switch *S* will not activate the alarm when required.

Redesign the circuit to eliminate this failure mode by using a domain-independent method.

Exercise 13.32

Loss of air or hydraulic pressure in the brake line represents a critical failure mode for trains. The consequence is a complete loss of braking capability, which can result in a catastrophic runaway accident. Using a domain-independent method, propose a solution that prevents this failure mode.

Exercise 13.33

An object (1) with mass *M* hangs from a pulley (2) using a steel band (3) (Figure 13.17). The steel band is secured by two types of screws: Two screws of variety *A* on the left-hand side, and two screws of variety *B* on the right-hand side. On each side, only one of the screws is sufficient to support the weight, the other screw is redundant. By using a domain-independent method, improve the reliability of the assembly of screws used for supporting the weight.

Exercise 13.34

Conventional circular saws, with teeth distributed uniformly around the circumference of the blade, tend to generate excessive noise. Using a domain-independent approach, propose a design modification that will reduce the noise.

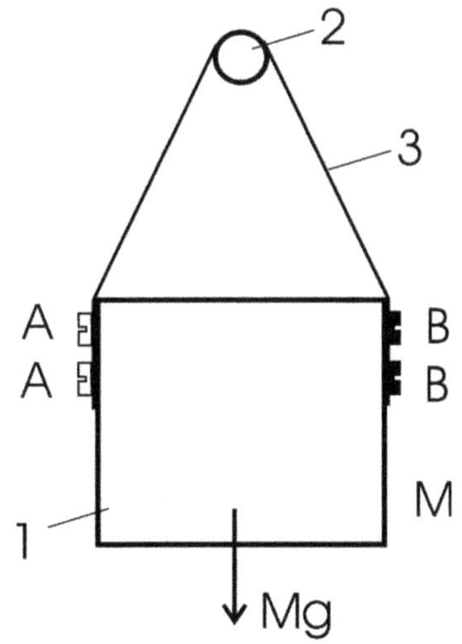

FIGURE 13.17 An assembly used for supporting weight.

Exercise 13.35

In the electronic circuit shown in Figure 13.18, the alternating current flowing through the wires from the power source generates an electromagnetic field that induces noise and degrades the circuit's performance. By using a domain-independent method, reduce the noise induced by the electromagnetic field.

Exercise 13.36

This exercise is adapted from Salamatov (1999).

Ultrasonic degassing of molten alloys requires a waveguiding rod that functions as an ultrasonic conductor. The primary failure mode arises from erosion of the rod caused by the to high-frequency oscillations, which release degradation products that contaminate the melt. How can this failure mode be mitigated through a domain-independent approach?

Exercise 13.37

In the clevis assembly from Figure 13.9, the blade (2) frequently experiences tensile and shear failures in the area of the pin (3) while the clevis (1) never experiences failures of the prongs. By using a domain independent method propose a solution which reduces the failure frequency of the blade (2) without increasing the volume of material used for manufacturing the clevis and without replacing the material of the blade and the clevis.

Exercise 13.38

This exercise is adapted from Orloff (2006).

The crowd-control fence in Figure 13.19 can be easily overturned by the forces exerted by people pushing against it. By using a domain-independent method for improving reliability, eliminate the possibility of overturning the fence.

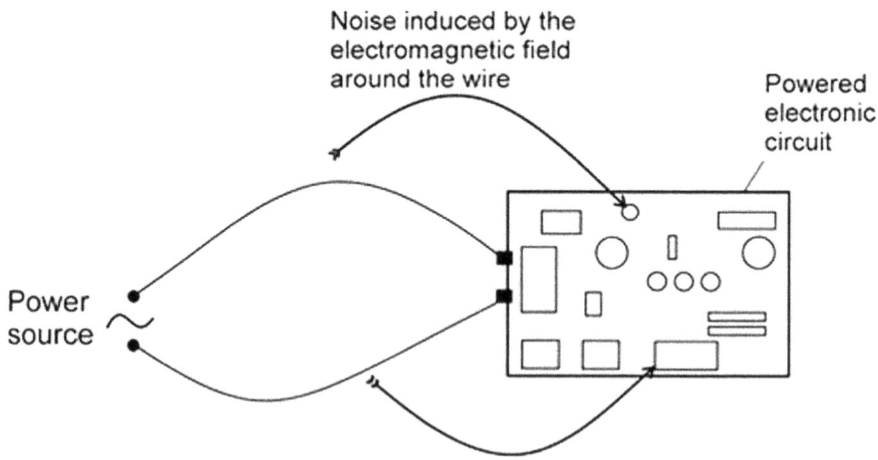

FIGURE 13.18 Electronic circuit powered by a source of alternating current.

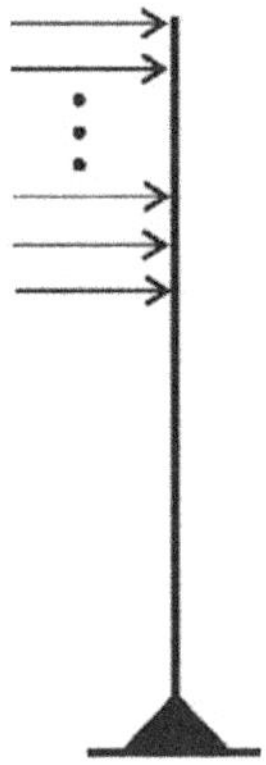

FIGURE 13.19 A crowd control fence with forces exerted by people pushing against it.

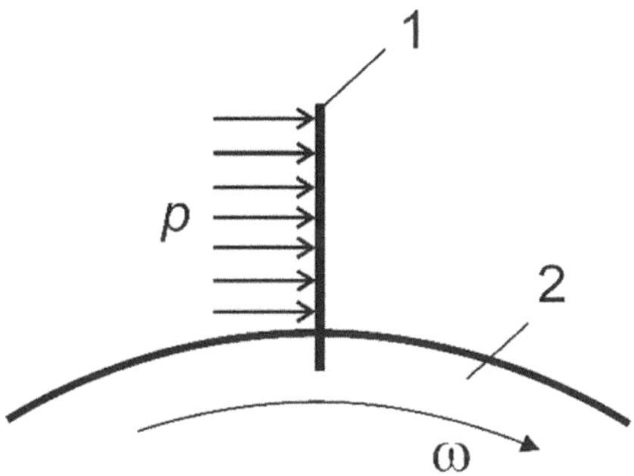

FIGURE 13.20 A rotating turbine wheel with a blade attached to it.

Exercise 13.39

This example is adapted from Pahl et al. (2007). The pressure p on the turbine blade (1) creates large bending stresses at the base of the blade in the attachment region of the turbine wheel (2) rotating with angular velocity ω (Figure 13.20). By using a domain-independent reliability improvement method, reduce the bending stresses at the base of the blade (and the risk of overstress failure).

Exercise 13.40

This exercise is adapted from (Terninko et al., 1998).

During the transportation of jet engines and electrical machinery with heavy rotors mounted on ball bearings (Figure 13.21), a critical failure mode has been identified. The jolting and shocks from rail joints and uneven roads cause dents in the bearing races. Use a domain-independent method to reduce the likelihood of dents in the bearing races.

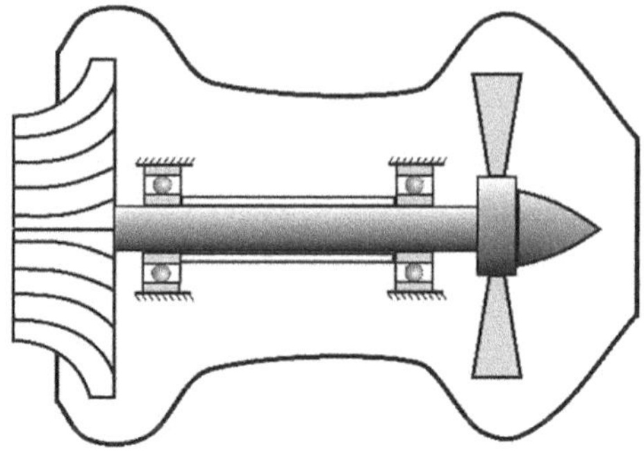

FIGURE 13.21 A jet engine with bearing races that are damaged during transportation.

Exercise 13.41

This exercise is adapted from Altshuller (1996).

The liquid dispenser shown in Figure 13.22 operates on a simple principle. When the liquid level reaches a certain point, the moment generated by the liquid about the pivot exceeds the counter-moment, causing the dispenser to tilt to the left and discharge the liquid. Once part of the container to the left of the pivot becomes lighter, the dispenser automatically begins returning to its initial position.

However, this device has a failure mode. As the left side becomes lighter during discharge, the dispenser enters an equilibrium state before the liquid has been fully released. By using a domain-independent method, redesign the device to eliminate this failure mode.

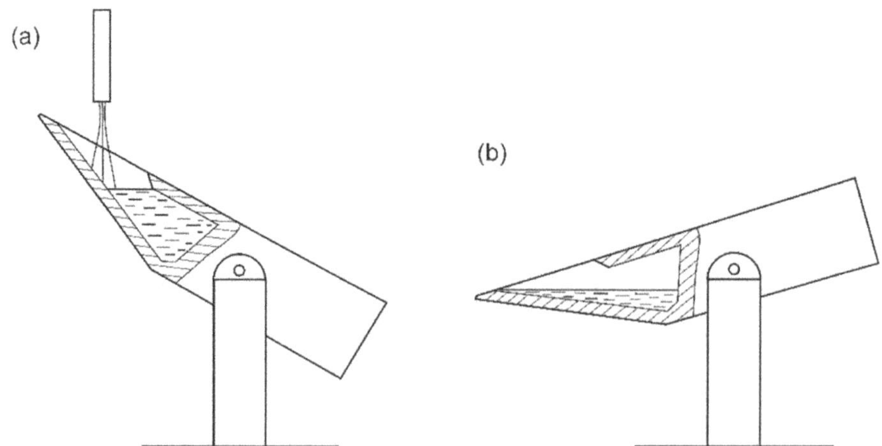

FIGURE 13.22 (a) Filling the dispenser with liquid. (b) The dispenser enters an equilibrium state before the liquid has been fully released.

Exercise 13.42

This exercise is adapted from Terninko et al. (1998). A filter consists of a tube lined with a porous material (Figure 13.23). An air–dust mixture flows through the tube, and dust particles become trapped in the porous lining. Due to the filter's design, the trapped dust cannot be dislodged or removed by simply passing clean air in the opposite direction. Propose a method to clean the filter and minimize process downtime associated with filter maintenance.

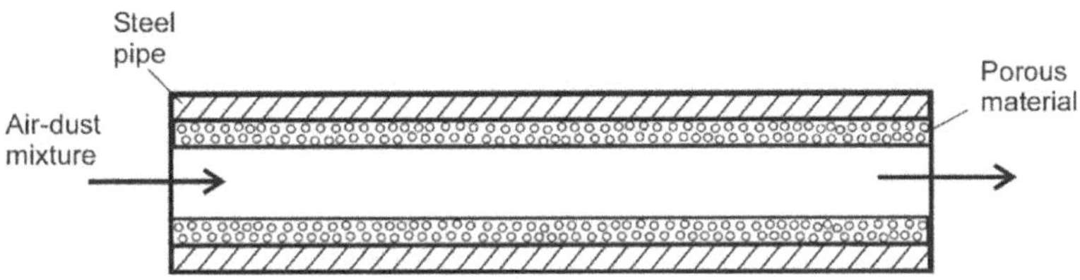

FIGURE 13.23 A filter capturing dust particles.

14 Solutions to the Tutorial Exercises from Chapter 13

Exercise 13.1
Thirteen failure modes have been identified for the clevis joint assembly in Figure 14.1. Preventing all 13 failure modes requires specifying the controllable design variables within the intersection of the domains of the design variables that prevent each listed failure mode.

Shear failure modes:

- Shear failure of pin 5;
- Shear failure of eye 2;
- Shear failure of clevis 4.

Compressive failure modes:

- Compressive failure of pin 5 due to excessive bearing pressure from the eye 2;
- Compressive failure of pin 5 due to excessive bearing pressure from the clevis 4;
- Compressive failure of clevis 4 due to excessive bearing pressure from the pin 5;
- Compressive failure of eye 2 due to excessive bearing pressure from the pin 5.

Tensile failure modes:

- Tensile failure of the blade in zone 1, away from the eye 2;
- Tensile failure in zone 3, away from the clevis prongs 4;
- Tensile failure of the blade around the eye 2;
- Tensile failure of the clevis prongs 4 in the vicinity of the hole.

Other failure modes:

- Bending of pin 5;
- Failure of clip 6.

Exercise 13.2
The root causes leading to the system failure mode 'leakage from an O-ring' are given in the Ishikawa diagram from Figure 14.2.

The mechanism behind the contributing factor 'too small clearance B-d' is illustrated in Figure 14.3. An O-ring with adequate clearance 'B-d' is shown in Figure 14.3a. Its proper functioning is depicted in Figure 14.3b. In contrast, Figure 14.3c demonstrates failure of an O-ring to seal properly when the clearance 'B-d' is too small.

 DOI: 10.1201/9781003679301-14

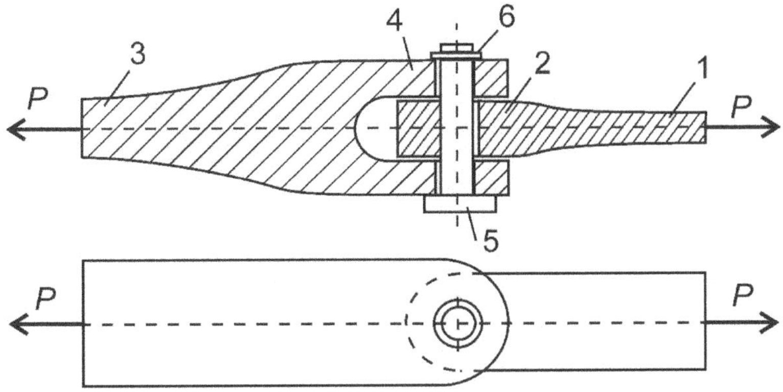

FIGURE 14.1 A clevis joint loaded with a constant force *P*.

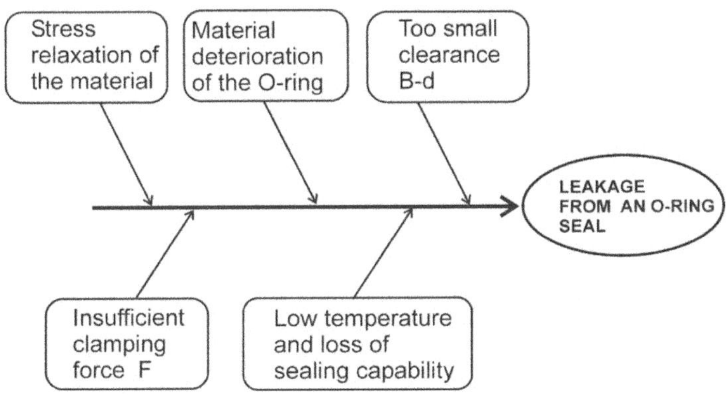

FIGURE 14.2 Ishikawa diagram for the failure mode 'leakage from an O-ring'.

Exercise 13.3

Apply the method of substitution. Replace mechanical cleaning with ultrasonic cleaning. Ultrasonic cleaning is an effective and non-destructive method for removing dirt and grease from fracture surfaces. The fractured spring segment is placed in a basket or suspended using a hook to avoid direct contact with the walls of the ultrasonic cleaning tank. It is then submerged in a suitable cleaning solution within the tank. After the ultrasonic cleaning, the component is thoroughly rinsed with clean distilled water to remove any residual cleaning agents. If necessary, the part is dried using a hot air dryer to prevent water spots or corrosion.

Exercise 13.4

Apply the method of substitution. Replace the electrical strain gauges with optical strain gauges. Optical strain gauges operate without physical contact, which eliminating failure modes related to poor bonding or surface imperfections. They also

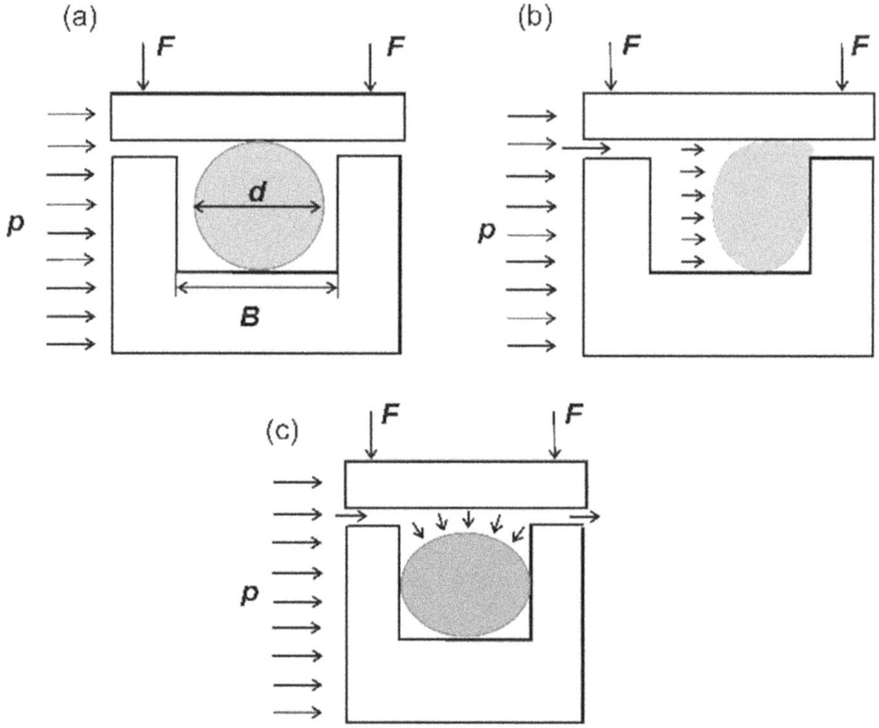

FIGURE 14.3 (a) An O-ring with clearance ‘B-d’; (b) operation of the O-ring with an adequate clearance B-d; (c) operation of the O-ring with too small clearance ‘B-d’.

eliminate measurement errors associated with surface preparation, greatly reducing variability. Additionally, optical strain gauges offer a significantly faster measurement in both laboratory and field applications.

Exercise 13.5

Apply the method of substitution. Replace traditional mechanically based detection methods with an acoustic method. The proposed solution involves using acoustic sensors to detect fluid or gas leaks. This approach offers the advantage of identifying micro-leaks and provides non-invasive, real-time monitoring.

The specific method could employ a combination of a microphone and a sound amplifier to detect characteristic hissing sounds indicating gas escaping from welds or joints.

Exercise 13.6

Apply the method of substitution. Replace mechanically based methods with the use of an electrostatic field. An electrostatic field is generated by connecting the light bulb to a high-voltage power source. The field strength is gradually increased until a corona discharge is detected around the filament of the light bulb. If this discharge occurs at a lower voltage than that of standard, non-defective bulbs, it indicates excessive internal pressure. The light bulb is then classified as defective and removed from production.

Exercise 13.7

Apply the method of separation (separation of properties). According to the principle of strength separation based on localized stress, the hinges in the original design in Figure 14.4a should be fabricated from material with significantly higher fatigue resistance than the rest of the container. This is demonstrated in the improved design from Figure 14.4b, where the hinges are made of fatigue-resistant material. This targeted separation of properties significantly enhances the overall reliability of the component by ensuring that each part is optimally equipped to resist the specific stresses it encounters.

Exercise 13.8

In the original design (Figure 14.5a), the pulling force F excites shear stresses $\tau_1 = F/(2\pi rt)$ along the surface of the cylinder (marked by dashed lines). Inserting a circular plate 3 of material with superior strength provides a separation of properties.

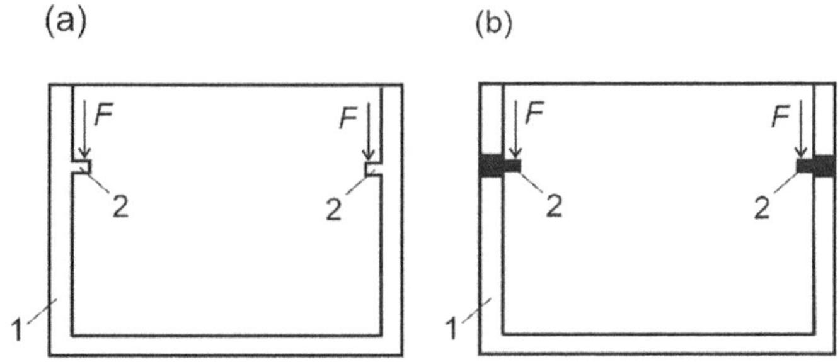

FIGURE 14.4 Improving the reliability of a container with rotating lid.

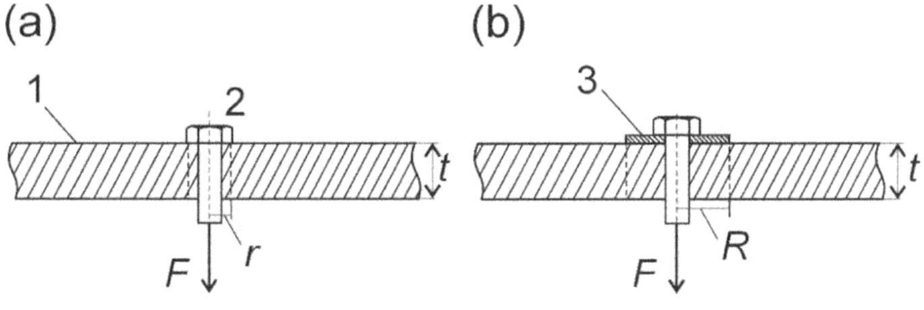

FIGURE 14.5 Improving the reliability of a connection through a low-strength component.

Because of the superior strength of plate 3, shear failure through the plate is excluded. The excited shear stress $\tau_2 = F/(2\pi Rt)$ is now along the surface of a cylinder (marked by dashed lines) with much larger radius ($R > r$) and therefore the shear stress $\tau_2 < \tau_1$ is significantly smaller compared to the shear stress characterizing design (a). The reliability of the connection has been improved by separation of properties, at a low cost.

Exercise 13.9

Apply logical separation. To prevent the operator's hand from entering the cutting zone of the guillotine, the machine is designed so that the cutting action is only triggered by the simultaneous activation of two separate handles/buttons, one for each hand. This design ensures that both operator's hands are engaged and away from the cutting area during activation. Thus, it is logically impossible for the operator's hand to be in the danger zone at the time of cutting. This solution is an example of a logical separation that significantly reduces accident risk.

Exercise 13.10

Apply separation of functions. The hybrid joint in Figure 14.6 combines an adhesive bond (1) with a mechanical fastener (2). In this configuration, a functional separation has been applied: the adhesive reduces stress concentration along the joint, while the mechanical fixing enhances peel resistance and increases overall joint stiffness.

Exercise 13.11

Apply separation of properties. The solution involves designing a pocket that remains continuously lined up with pellets (Figure 14.7). As a result, the stream of pellets is separated from bend and no longer strikes the pipe directly. Instead, the pellets impact the blanket of pellets in the pocket, which effectively eliminates wear.

Exercise 13.12

Apply separation in geometry. The design protecting against overtightening the screw can be achieved by using a screw with a shoulder, as shown in Figure 14.8.

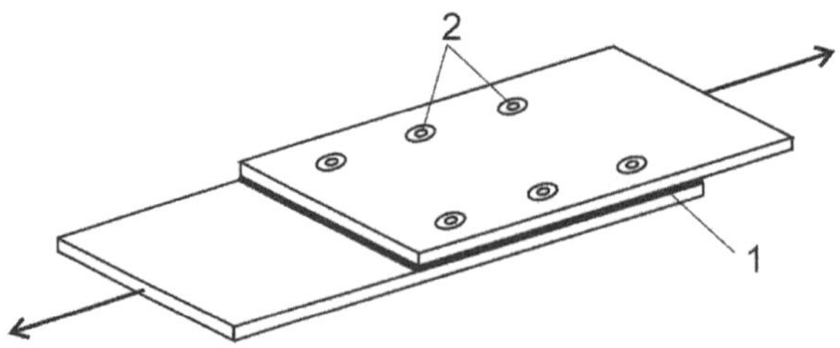

FIGURE 14.6 Separation of functions to compensate mutual deficiencies.

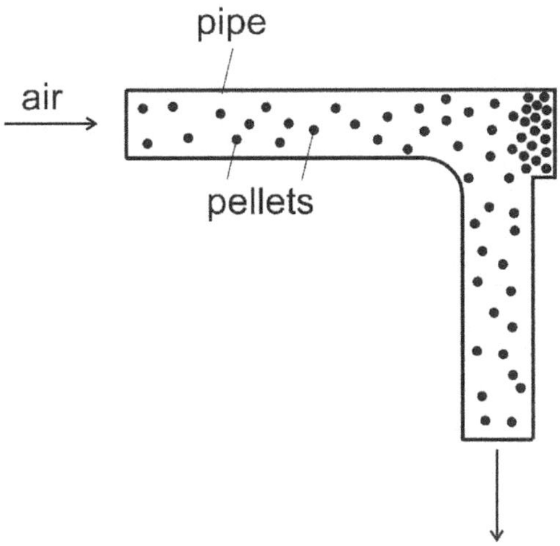

FIGURE 14.7 A pipe transporting pellets, with a designed pocket.

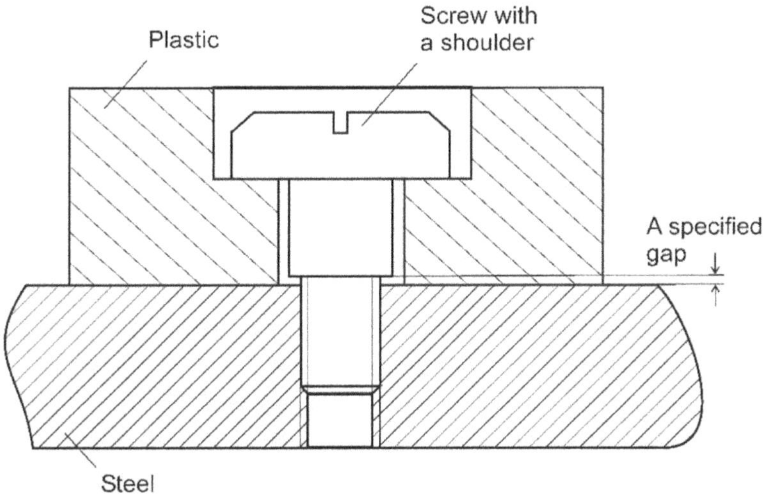

FIGURE 14.8 Fastening a plastic sheath with a shouldered screw.

Exercise 13.13

The design solution, as illustrated in Figure 14.9, is achieved by applying separation in geometry. In the original design shown in Figure 14.9a, the cylindrical pin (3) is replaced by a conical pin (3) in the modified design (Figure 14.9b). The tapering angle of the conical pin is chosen to be greater than the angle of friction, allowing for easy disassembly. This design modification eliminates the clearance between pin (3), clevis (1), and eye (2), thus removing the relative motion between these components and significantly reducing mechanical friction. As a result, wear caused by the pulsating force F is reduced significantly.

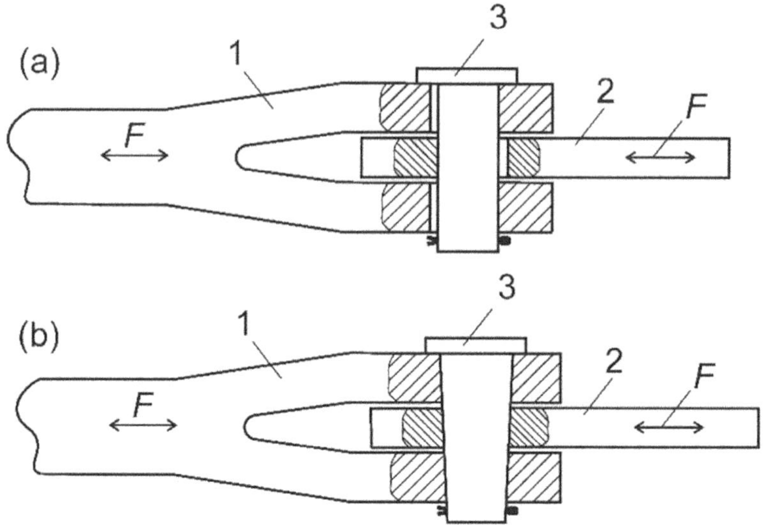

FIGURE 14.9 (a) Original and (b) modified design of the clevis joint subjected to a pulsating force.

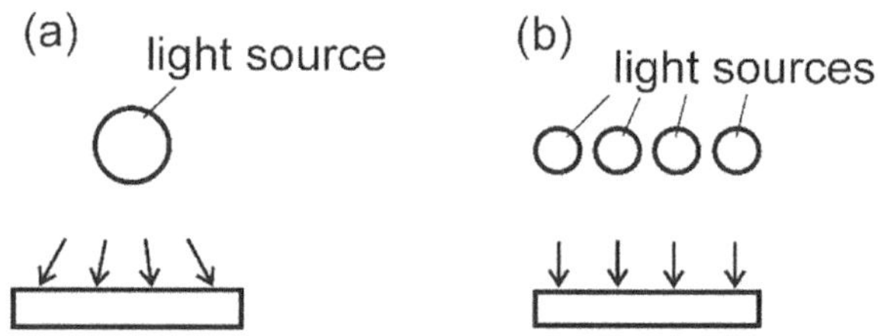

FIGURE 14.10 Segmenting (a) a single light source into (b) multiple light sources.

Exercise 13.14

Apply the method of segmentation. If the single light source (Figure 14.10a) is segmented into multiple smaller light sources (Figure 14.10b) with the same total luminous flux, failure of a single light source or even failure of several light sources will not be catastrophic and will not entail a shutdown of the manufacturing process. Segmentation reduces risk by replacing a single critical failure at a macro level with gradual deterioration of the system at a micro-level through non-critical failures.

Exercise 13.15

Apply the method of segmentation. A potential solution is to segment the single bolt connecting the hitch receiver (1) and the trailer hitch (2) into two separate bolts, labelled (3) and (4) in Figure 14.11. During normal operation, only bolt (4) bears the stresses induced by the trailer load F. In the event of bolt (4) failure, the load is automatically transferred to bolt (3), allowing continued control of the trailer.

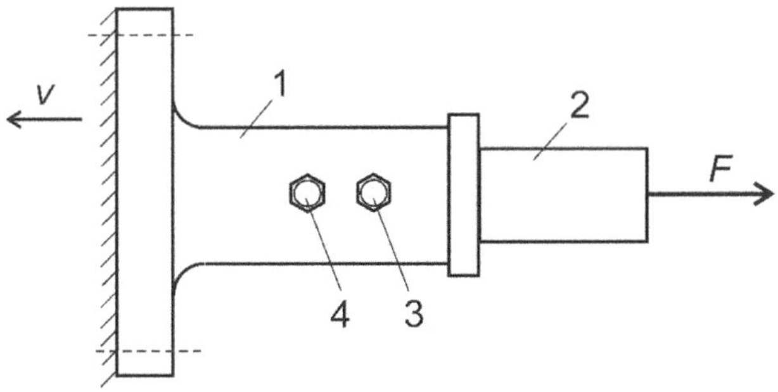

FIGURE 14.11 A trailer hitch assembly with a significantly enhanced safety.

Exercise 13.16

Apply the method of segmentation. The single electrical bus supplying power is divided into multiple independent power buses, each of which is powered by different generators. The reliability benefit from segmentation is the fault isolation. Critical systems are connected to multiple buses, ensuring continued operation even after partial system failures. Faults can also be diagnosed and repaired within specific segments without taking down the whole system.

Exercise 13.17

Apply the method of segmentation. Design the toolheads, arms, and the mobility bases to be modular and easily replaceable.

Exercise 13.18

Apply the method of segmentation. Segmenting the brittle column into bricks significantly enhances fault tolerance against critical flaws. In a segmented column (Figure 14.12), each individual brick can withstand a much higher density of critical flaws, compared to a monolithic structure. Because of the relatively weak interfaces between the bricks, a crack starting from a critical flaw in any of the bricks cannot spread through the neighbouring bricks and trigger the collapse of the entire column. A crack appearing in a particular brick causes only failure of that brick. As a result, localized damage, such as a crack in a single brick, remains contained, thus preventing the crack from propagating throughout the entire column.

Exercise 13.19

Apply the method of segmentation. A spline connection (Figure 14.13b) instead of key connection (Figure 14.13a) can be used. Splines can be viewed as segmented keys that are uniformly spaced around the shaft. This type of segmentation distributes the load more evenly, reducing stress concentrations and improving the assembly's resistance to fracture in case of overload.

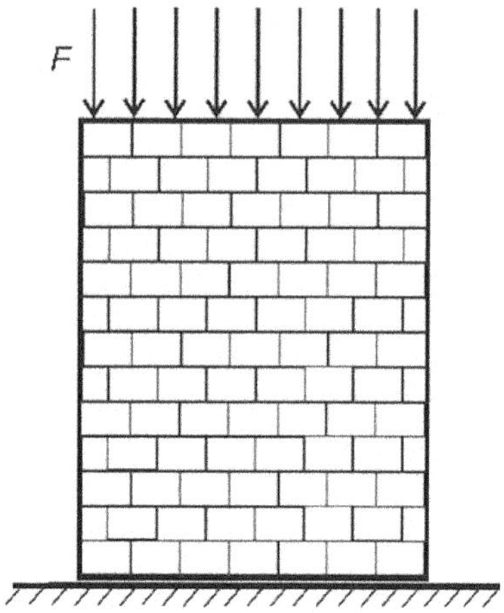

FIGURE 14.12 Segmentation improves the fault tolerance of a brittle column loaded in compression.

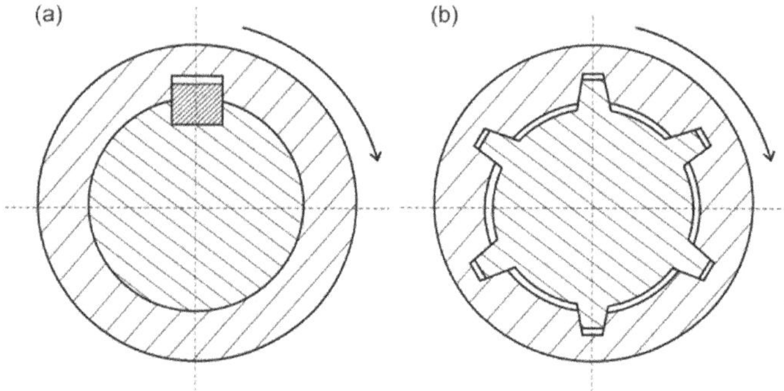

FIGURE 14.13 (a) A key assembly and (b) a spline assembly; splines can effectively be regarded as segmented keys uniformly spaced around the shaft.

Exercise 13.20

Apply the segmentation method: use multiple hoist ropes attached to the same hook or lifting mechanism. The ropes should be rated such that if any of the ropes fails, the remaining ropes can immediately share the load preventing dropping of the hoisted object and providing built-in redundancy.

In some systems, ropes pass over sheaves with tension equalizers to ensure even load distribution. Additionally, shock absorbers or spring systems can dampen sudden load changes, preventing overload on the remaining ropes.

These measures eliminate single-point failures, ensuring that a rope break does not result in dropping the hoisted object.

Exercise 13.21

Segmentation at the atomic level can be applied to ensure that microelements are distributed evenly across the soil or foliage. Rather than relying on dry blending, liquid fertilizers containing copper and manganese can be used. These fertilizers may be applied directly to the soil, sprayed onto foliage, or introduced into irrigation water, providing far greater uniformity than dry blending. For example, irrigation water can be passed through an electrolyzer equipped with electrodes made from a manganese–copper alloy (Salamatov, 1999). This process delivers strictly proportioned amounts of microelements, with dosing achieved at the atomic level.

Exercise 13.22

Apply the segmentation method. The jaws of the vice can be segmented by placing multiple hardened bushings between the flat surface of the vice and the workpiece (Figure 14.14). Each bushing is free to move independently, allowing it to adapt to the complex surface geometry of the workpiece.

Exercise 13.23

Introduce a deliberate weakness. A shear pin can be incorporated to create a deliberate weakness between the door handle and the locking mechanism. In the event of a forceful slam, the shear pin will fail, disconnecting the force transmission from the handle to the locking mechanism and preventing damage.

Exercise 13.24

Apply the method of deliberate weakness together with the method of segmentation. To reduce wear of the conveyor belt, install replaceable steel segments fitted with rubber wear pads on top. The rubber pads absorb most of the impact and friction from the polishing heads, thus protecting the underlying belt.

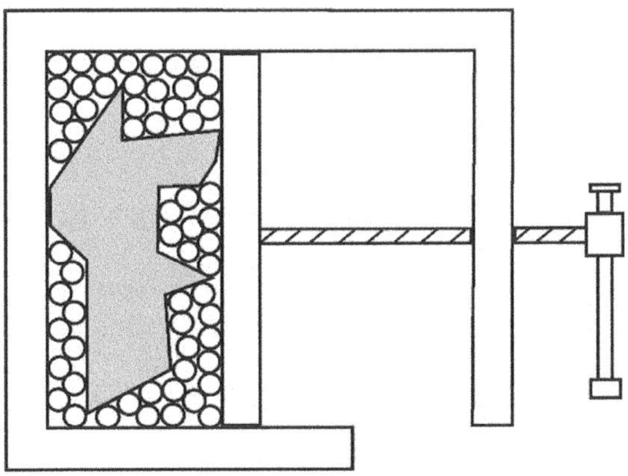

FIGURE 14.14 A vice with segmented jaws capable of securing workpieces with complex shape.

Regular inspections should be carried out. When a worn pad is identified, the corresponding steel segment can be removed, the damaged pad replaced and the segment reinstalled.

This solution is cost-effective, as replacing a small wear pad is far less expensive than replacing the entire conveyor belt. Consequently, the belt's service life is significantly extended. Furthermore, worn segments can be replaced quickly during scheduled maintenance, simplifying upkeep and minimizing unplanned downtime.

Exercise 13.25

Apply the method of deliberate weaknesses. Incorporate a circuit breaker to prevent overload-related failures. The circuit breaker could operate in three distinct states: 'Closed', corresponding to normal operation where all requests are allowed through; 'Open', where requests are immediately rejected to prevent further strain and 'Half-open', where a limited number of requests are permitted.

The half-open state could implement priority-based request queuing, allowing high-priority traffic while dropping or deferring low-priority requests. This ensures that critical operations proceed during partial recovery, while minimizing the risk of repeated overload.

Exercise 13.26

Apply the method of deliberate weakness and incorporate a *torque limiter* (also known as a slip clutch or overload clutch), which is a mechanical device designed to disconnect or slip when the torque exceeds a preset threshold. This prevents the transmission of damaging loads and protects the drive mechanism from mechanical failure.

By automatically disengaging or slipping under excessive load, the torque limiter safeguards critical components such as gears, motors, and joints, thereby avoiding catastrophic damage.

The process begins by determining the maximum safe torque the robotic arm's drive can withstand. A torque limiter rated slightly below this threshold is then selected and installed between the motor and the driven component (e.g., gear or joint).

Exercise 13.27

These failure modes can be mitigated or eliminated applying an *energy-absorbing deliberate weakness* (Figure 14.15).

Introducing an elastic component (4) between the base (5), to which the wheel (1) is attached and part (6) attached to the suitcase, results in significantly reduced shock loads on the suitcase body (see Figure 14.15). Part (2) connects to the suitcase via part (6) and pivots around the pin (3). The elastic component (4) functions as a buffer, absorbing impact energy from the wheel and reducing the risk of mechanical damage to both the wheel and the wheel mount.

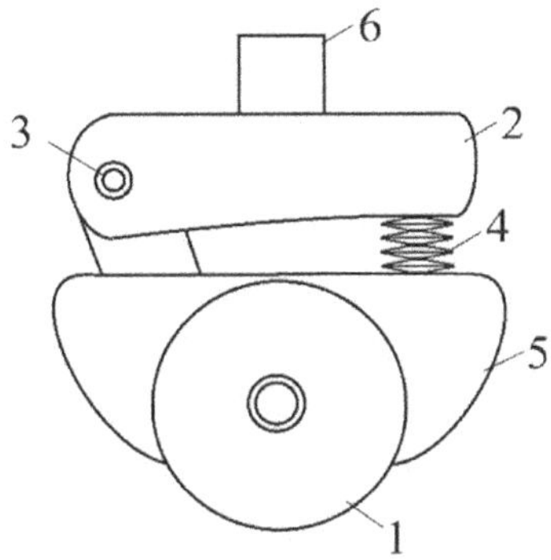

FIGURE 14.15 Design of a luggage wheel implementing a deliberate weakness absorbing impact energy.

Exercise 13.28
Incorporating a deliberate weakness can be implemented through an elastic insert within the vessel. As water freezes inside the vessel, the expanding ice exerts pressure on the elastic insert, causing it to compress and accommodate the ice expansion while the vessel itself remains undamaged.

Exercise 13.29
Apply the method of inversion. A reinforcing effect can be achieved by inverting the position of the component relative to its support, as illustrated in Figure 14.16. In the configuration shown in Figure 14.16a, the brackets support a component subjected to a varying force F. The resulting effective stress range $\Delta\sigma_1$ in the supporting brackets is also shown in Figure 14.16a.

However, by inverting the relative positions of the component and the supporting brackets, as depicted in Figure 14.16b, the stress range can be significantly reduced. This inverted arrangement results in a smaller stress range $\Delta\sigma_2$. The smaller stress range leads to extended fatigue life (Paris et al., 1961).

Exercise 13.30
Use the inversion method. Fill the container with pressurized air and submerge it in water. Air bubbles escaping from leaks are much easier to detect than traces of liquid seeping out.

Exercise 13.31
The failure mode in the original circuit can be eliminated by applying the method of inversion. Convert the normally open switch S and the normally open reed relay contact K into a normally closed switch $\overline{S}$ and a normally closed reed relay contact $\overline{K}$, as shown in Figure 14.17a. The relay used in this configuration is a high-coil-resistance reed relay.

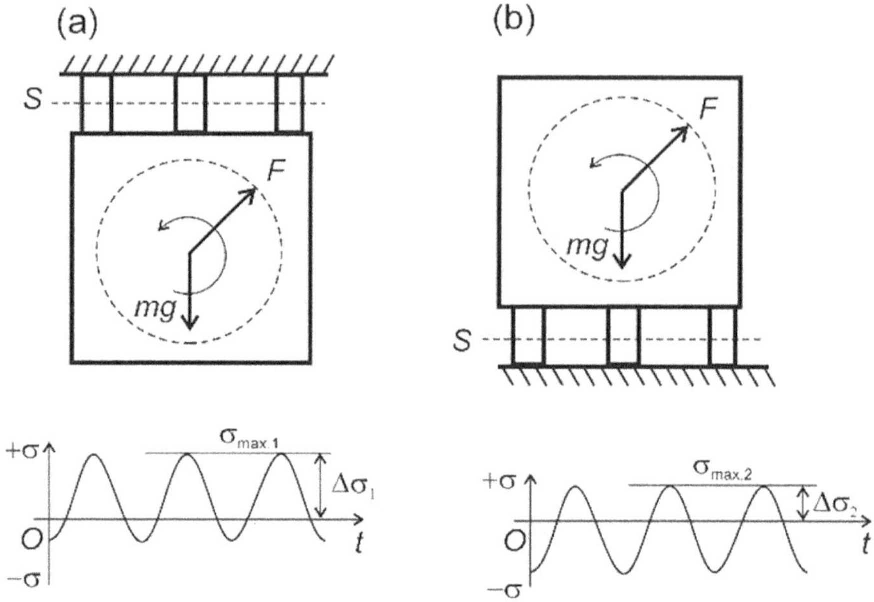

FIGURE 14.16 (a) Improving the reliability of the supporting brackets by (b) inverting the position of the component with respect to its support.

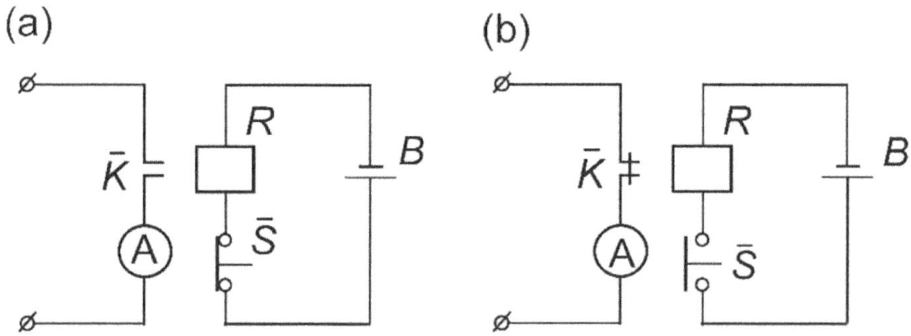

FIGURE 14.17 Improving the reliability on demand of an alarm system by inversion.

In the modified circuit, when the normally closed switch is in its default (closed) state, the reed relay coil R is energized. This keeps the normally closed relay contact open, preventing the alarm from being activated. However, if the switch is opened, as illustrated in Figure 14.17b, the coil becomes de-energized, the reed contact closes, and the alarm is activated.

The alarm will be triggered under any of the following conditions:

- Pressing the normally-closed switch $\bar{S}$;
- An open circuit caused by a fault;

- An open circuit caused by deliberate wire cutting;
- A depleted (flat) battery *B*.

In this configuration, a false alarm is considered a preferable outcome compared to the failure to activate the alarm when needed. In fact, false alarms serve a useful purpose by signalling that the alarm circuit requires maintenance.

Thanks to the high resistance of the reed relay coil *R* (typically several kilo-ohms), the power required to keep the reed contact open is negligible.

By applying the method of inversion in this example, system security is enhanced without substantial additional cost.

Exercise 13.32

Use the method of inversion. Design the service brakes to be spring-applied and pressure-released, so that pressure is used only to hold the brakes in the released position. In this configuration, any loss of air or hydraulic pressure automatically engages the brakes.

Exercise 13.33

Apply the method of asymmetry. The reliability network related to the supporting screws of the original assembly is given in Figure 14.18a. It is a series-parallel network. The reliability R1 of the system of screws can be improved by an asymmetric arrangement of the redundant screws. This method works irrespective of the actual reliabilities of the screws. The new assembly with improved reliability R2 (R1 < R2) is presented in Figure 14.18b.

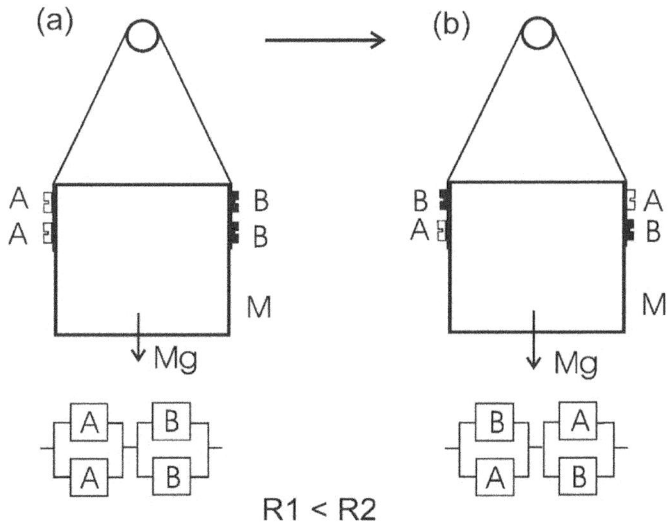

FIGURE 14.18 (a) Original assembly supporting a load with mass *M* and (b) an assembly with enhanced reliability.

Exercise 13.34

Noise in circular saws can be mitigated by exploiting the asymmetry in spacing the teeth. Uniform tooth spacing leads to constructive interference of the generated acoustic waves, thereby inducing resonance and amplifying noise. In contrast, asymmetric tooth spacing, achieved by arranging teeth at unequal angular intervals, disrupts the periodicity of excitation forces. This interruption suppresses resonant vibrations and reduces acoustic emissions.

Exercise 13.35

Apply the method 'increasing the level of balancing'. Use twisted-pair wiring. Twisted conductors carry equal and opposite currents, causing their electromagnetic fields to cancel each other out. This configuration inherently provides self-balancing. (See Figure 14.19.)

Exercise 13.36

Increasing the level of balancing may be applied. The waveguiding rod, which serves as a ultrasonic conductor, should be made of the same alloy as the melt. This prevents contamination, since erosion of the rod under high-frequency oscillations will no longer release foreign degradation products into the melt.

Exercise 13.37

Apply the method of increasing the level of balancing. The frequent tensile and shear failures of the blade in the area of the pin occur because the stresses in the blade are significantly higher than those in the prongs. The internal stresses in the clevis assembly can be rebalanced without adding extra material by reducing the thickness of the prongs while increasing the thickness of the blade in the area of the pin. The design should then be checked to make sure that no tensile, shear, and compressive failure modes will be associated with the prongs.

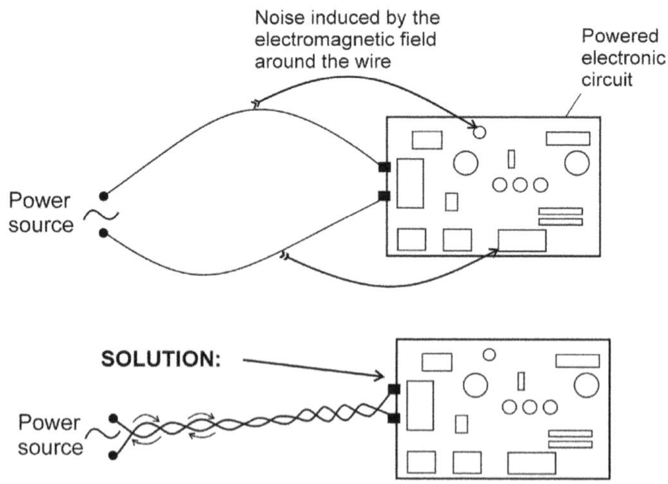

FIGURE 14.19 Reducing the noise induced by an electromagnetic field by twisting wires.

Exercise 13.38

Apply the method of self-reinforcement. Self-reinforcing design principles can be effectively applied to improve the stability of crowd-control fences and eliminate the risk of overturning. A conventional fence without self-reinforcement (Figure 14.20a) can be easily toppled by the horizontal forces exerted by individuals pushing against it. However, when the lower edge of the fence on the crowd-facing side is extended or widened (Figure 14.20b), people pushing against the fence must stand on this extended base. As a result, their own body weight generates a stabilizing moment M' that opposes the overturning moment M created by the pushing forces.

The more people push, the greater the counteracting force becomes, transforming the destabilizing action into a self-reinforcing stabilizing effect.

Exercise 13.39

A self-reinforcing solution based on capturing rotational motion can be applied. The bending stresses (undesirable outcome) at the base of the blade (Figure 14.21a), caused by the tangential forces from the pressure, can be reduced if the centrifugal force *F*, generated by the blade's rotation, is directed to produce an opposing moment. This opposing moment counteracts the moment induced by the tangential forces. Such redirection of the centrifugal force (resulting in a proportional compensating factor) can be achieved by inclining the blades at an angle relative to the vertical axis (Figure 14.21b).

In the design shown in Figure 14.21b, the centrifugal force *F* creates a proportional moment that opposes the bending moment from the tangential forces. This significantly reduces the overall bending moment and the stresses at the blade base. The higher the angular velocity ω, the greater the centrifugal force (which is proportional to the square ω^2 of the angular velocity), and the stronger the counteracting moment it generates.

As a result, the assembly becomes more capable of withstanding sudden increases in pressure, which enhances its resilience and reduces the risk of failure. This

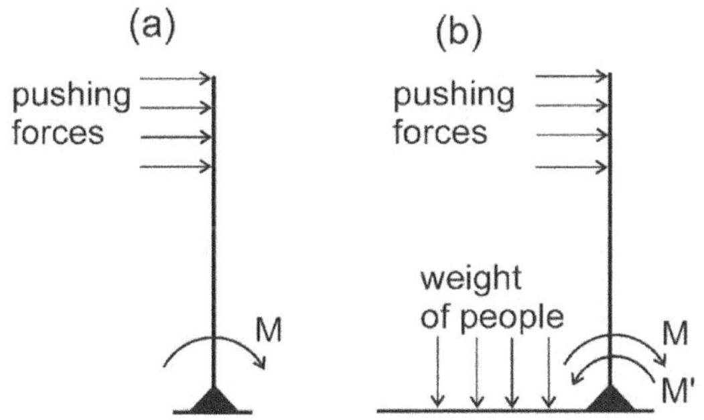

FIGURE 14.20 (a) Original crowd control fence; (b) self-reinforcing solution for the crowd control fence.

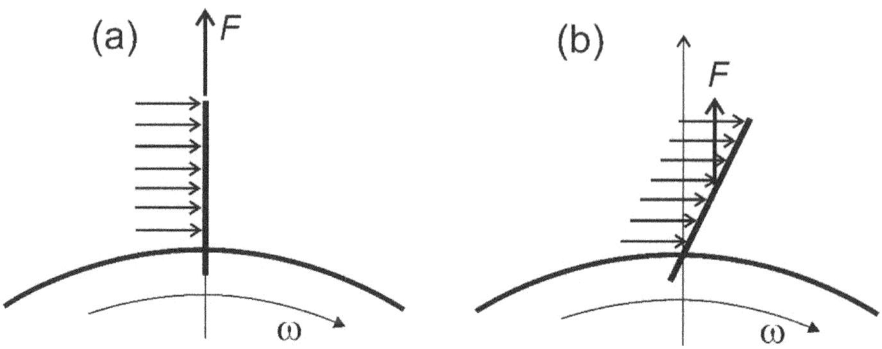

FIGURE 14.21 Reducing the stresses at the base of a blade attached to a rotor by capturing a proportional compensating moment (a) without self-reinforcement; (b) with a self-reinforcement.

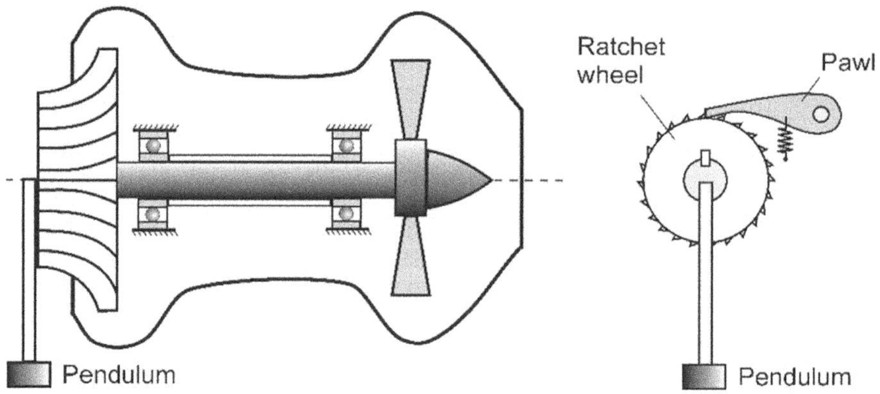

FIGURE 14.22 Preventing the bearing races of a jet engine through self-service provided by a pendulum and ratchet mechanism.

improvement in reliability through self-reinforcement is achieved without additional cost.

Exercise 13.40

Denting of the bearing races can be prevented through self-reinforcement by self-service. The energy from shocks is harnessed to continuously rotate the rotor using a pendulum and ratchet mechanism (see Figure 14.22). This slow, continuous rotation distributes the load across different points on the bearing race, preventing damage from accumulating in a single location.

Exercise 13.41

Apply the principle of self-reinforcement by using gravity as a compensating factor. The dispenser can be hollowed out on the opposite side and a heavy ball can be inserted into the blind hole, as illustrated in Figure 14.23 (Altshuller, 1996). When the dispenser tilts, the ball shifts toward the pivot, significantly reducing the returning moment. As a result, the scoop continues to tilt, since no sufficient counterbalancing moment remains to retain the liquid. This allows the scoop to release its entire liquid content.

Exercise 13.42

The principle of self-service can be applied to enable quick filter cleaning and minimize downtime. In this method, the sealed filter in Figure 13.23 is exposed to a high pressure of 5–10 atmospheres (Terninko et al., 1998). The pressure is then abruptly reduced to 1 atmosphere, causing the rapid release of air from the pores. This sudden pressure drop expels dust particles along with the escaping air.

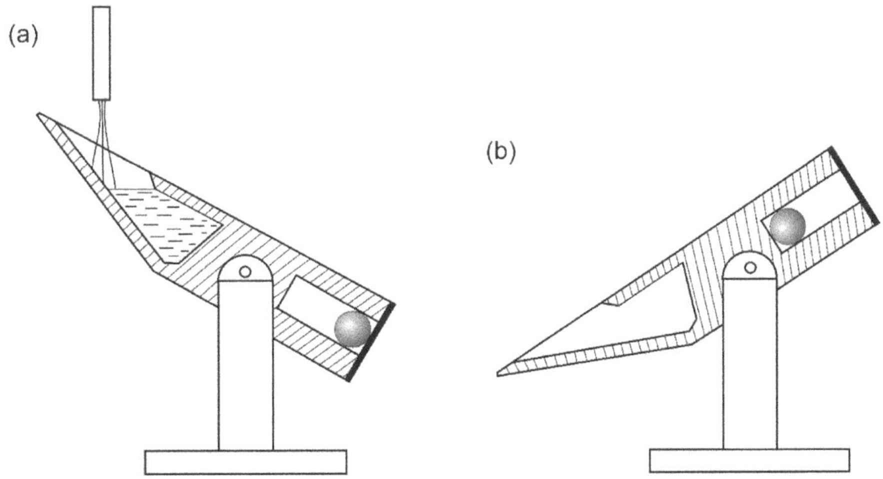

FIGURE 14.23 (a) Filling the dispenser with liquid; (b) emptying the dispenser. The displacement of the ball towards the pivot reduces the resisting counter-moment and helps to release the entire liquid content.

References

Altshuller G.S. (1984). *Creativity as an exact science: The theory of the solution of inventive problems*, Gordon and Breach Science Publishing, New York.

Altshuller G.S. (1996). *And suddenly the inventor appeared, TRIZ, the theory of inventive problem solving*, Translation from Russian by Lev Shulyak, Technical Innovation Center, Worcester, MA.

Altshuller G.S. (1999). *The innovation algorithm, TRIZ, systematic innovation and technical creativity*, Technical Innovation Center, Worcester, MA.

Ang A.H.S. and Tang W.H. (1975). *Probability concepts in engineering planning and design*, vol. 1, *Basic principles*, Wiley, New York.

Aqel I. and Mohamed Mellal M. (2023). Optimal reliability allocation of heterogeneous components in pharmaceutical production plant. *International Journal on Interactive Design and Manufacturing (IJIDeM)*, 17, pp.1711–1720.

Ashby M.F. and Jones K. (2002). *Materials and design: The art and science of material selection in product design*, Butterworth Heinemann, Oxford.

Aven T. (2003). *Foundations of risk analysis: A knowledge and decision-oriented perspective*, Wiley, Chichester.

Bakshi S., Dhillon P. and Maruvada T. (2014). Design and optimisation of planetary gearbox for a formula student vehicle. *SAE Internatonal Journal of Materials and Manufacturing*, 7(3), pp.688–695.

Barlow R.E. and Proschan F. (1965). *Mathematical theory of reliability*, Wiley, New York.

Barlow R.E. and Proschan F. (1975). *Statistical theory of reliability and life testing*, Rinehart and Winston, New York.

Bazovsky I. (1961). *Reliability theory and practice*, Prentice-Hall, Englewood Cliffs, NJ.

Beer F.P., Russell Johnston E. Jr. and DeWolf J.T. (2002). *Mechanics of materials*, 3rd ed., McGraw-Hill, New York.

Bendsøe M.P. and Sigmund O. (2003). *Topology optimization – Theory, methods and applications*, Springer Verlag, Berlin, Heidelberg.

Bergman B., de Mare J., Loren S. and Svensson T. editors (2009). *Robust design methodology for reliability*, Wiley, Chichester.

Bikas H., Stavropoulos P. and Chryssolouris G. (2016). Additive manufacturing methods and modelling approaches: A critical review. *International Journal of Additive Manufacturing Technology*, 83, pp.389–405.

Billinton R. and Allan R.N. (1992). *Reliability evaluation of engineering systems*, 2nd ed., Plenum Press, New York.

Bird G.C. and Saynor D. (1984). The effect of peening shot size on the performance of carbon-steel springs. *Journal of Mechanical Working Technology*, 10(2), pp.175–185.

Booker J.D., Raines M. and Swift K.G. (2001). *Designing capable and reliable products*, Butterworth-Heinemann, Oxford.

Budynas R.G. (1999). *Advanced strength and applied stress analysis*, 2nd ed., McGraw-Hill, New York.

Budynas R.G. and Nisbett J.K. (2015). *Shigley's mechanical engineering design*, 10th ed., McGraw-Hill, New York.

Buragohain M.K. (2017). *Composite structures: Design, mechanics, analysis, manufacturing, and testing*, CRC Press, Boca Raton, FL.

Burrell N.K. (1985). Controlled shot peening of automotive components. *SAE Transactions*, Section 3, 94(850365).

Campbel J. (2015). *Complete casting handbook, metal casting processes, metallurgy, techniques and design*, 2nd ed., Butterwoth-Heinemann, Amsterdam.
Childs P.R.N. (2014). *Mechanical design engineering handbook*, Elsevier, Amsterdam.
Ciupitu L. (2018). Adaptive balancing of robots and mechatronic systems. *Robotics*, 7, p.68.
Collins J.A. (2003). *Mechanical design of machine elements and machines*, Wiley, New York.
Coria I., Abasolo M., Gutiérrez A. and Aguirrebeitia J. (2020). Achieving uniform thread load distribution in bolted joints using different pitch values. *Mechanics & Industry*, 21, p.616.
Costache A., Glejbol K., Sivebak I.M. and Berggreen C. (2016). Improved friction joint with self-locking grips. *Journal of Offshore Mechanics and Arctic Engineering*, 138, p.8.
Crouhy M., Galai D. and Mark R. (2006). *The essentials of risk management*, McGraw-Hill, New York.
Cullen The Hon Lord. (1990). *The public inquiry into the piper alpha disaster*, HMSO, London.
Dasgupta A. and Pecht M. (1991). Material failure mechanisms and damage models. *IEEE Transactions on Reliability*, 40(5), pp.531–536.
DeGroot M. (1989). *Probability and statistics*, Addison-Wesley, Reading, MA.
Dhillon B.S. (2017). *Engineering systems reliability, safety, and maintenance*, CRC Press, New York.
Dhillon B.S. and Singh C. (1981). *Engineering reliability: New techniques and applications*, Wiley, New York.
Ding Y., Hu Y. and Li S. (2021). Redundancy optimization for multi-performance multi-state series-parallel systems considering reliability requirements. *Reliability Engineering & System Safety*, 215, 107873.
Ebeling C.E. (1997). *An introduction to reliability and maintainability engineering*, McGraw-Hill, New York.
Eder W.E. and Hosnedl S. (2008). *Design engineering*, CRC Press, Boca Raton, FL.
Edwards B. and Cox C. (2002). Revolutionary concepts for helicopter noise reduction, NASA report, NASA/CR-2002-211650, 20020051150, May 2002.
Erhard G. (2006). *Designing with plastics*, Hanser, Munich.
Fantner G.E., Oroudjev E., Schitter G., Golde L.S., Thurner P., Finch M.M. and Hansma P.K., et al. (2006). Sacrificial bonds and hidden length: Unraveling molecular mesostructures in tough materials. *Biophysical Journal*, 90(4), pp.1411–1418.
Feng X., Ouyang M., Liu X., Lu L., Xia Y. and He X. (2017). Thermal runaway mechanism of lithium ion battery for electric vehicles: A review. *Energy Storage Materials*, 10, pp.246–267.
Florin L., Petru C. and Costin B. (2020). Optimization methods for redundancy allocation in large systems. *Vietnam Journal of Computer Science*, 7(3), pp.1–19.
French M. (1999). *Conceptual design for engineers*, 3rd ed., Springer-Verlag, London.
Fu H.M., Kang Y.M. and Shen H.G. (2010). Collection efficiency of fibrous filter on dust load. In *2010 4th International Conference on Bioinformatics and Biomedical Engineering* (pp.1–5). IEEE, Chengdu, China.
Fu Y., Ge E., Su H., Xu J. and Li R. (2015). Cold expansion technology of connection holes in aircraft structures: A review and prospect. *Chinese Journal of Aeronautics*, 28(4), pp.961–973.
Gadd K (2011). *TRIZ for engineers: Enabling inventive problem solving*, Wiley, Chichester.
Gere J. and Timoshenko S.P. (1999). *Mechanics of Materials,* 4th ed., Stanley Thornes, Cheltenham.
Giddens A. (1999). Risk and responsibility. *The Modern Law Review*, 62(1), pp.1–10.
Gu Y., Yan D., Verma V., Wang P., Stan M. and Zhang X. (2018) Exploiting read/write asymmetry to achieve opportunistic SRAM voltage switching in dual-supply near-threshold processors. *Journal of Low Power Electronics and Applications*, 8(3), p.28.

Gullo L.G. and J. Dixon. (2018). *Design for safety*, Wiley, Chichester.

Haddon Jr. W. (1973, August). Energy damage and the ten counter-measure strategies. *Human Factors Journal*, 15(4), pp.355–366.

Hale A., Goossens L., Ale B., Bellamy L., Post J., Oh J. and Papazoglou I.A. (2004). Managing safety barriers and controls at the workplace. In *Proceedings of PSAM7-ESREL'2004 Conference* (pp.609–613), 14–18 June, Berlin.

Hardy G., Littlewood J.E. and Pólya G. (1999). *Inequalities*, Mathematical Library, Cambridge University Press, Cambridge.

Hearn E.J. (1985). *Mechanics of materials, vol. 1 and 2*, 2nd ed., Butterworth-Heinemann, Oxford.

Henley E.J. and H. Kumamoto (1981). *Reliability engineering and risk assessment*, Prentice-Hall, Englewood Cliffs, NJ.

Hollnagel E. (2016). *Barriers and accident prevention*, Routledge, Abingdon.

Hoyland A. and Rausand M. (1994). *System reliability theory*, Wiley, New York.

HSE. (1992). *The tolerability of risk from nuclear power stations*, Health and Safety Executive, London.

Ilic O., Kaminer I., Lahini, Y., Buljan H. and Soljacic, M. (2016). Exploiting optical asymmetry for controlled guiding of particles with light. *ACS Photonics*, 3(2), pp.197–202.

Ionescu C. and Drumea A. (2019). Extended current range of active balancing and monitoring circuits for supercapacitor modules. In *2019 IEEE 25th International Symposium for Design and Technology in Electronic Packaging (SIITME)* (pp.366–371), 23–26 October, IEEE, Cluj-Napoca, Romania.

Jeon S., Yun J. and Bae1 S. (2015). Active cell balancing circuit for series connected battery cells. In *9th International Conference on Power Electronics-ECCE Asia June 1–5, 2015 / 63*, Convention Center, Seoul.

Kalpakjian S. and Schmid S. (2013). *Manufacturing engineering & technology*, 7th ed., Prentice Hall, Jurong, Singapore.

Kaplan S. and Garrick B.J. (1981). On the quantitative definition of risk. *Risk Analysis*, 1, pp.11–27.

Kaplan, S., Visnepolschi S., Zlotin B. and Zusman A. (1999). *New tools for failure and risk analysis: Anticipatory falure determination*, Ideation International, Detroit.

Kelly P.M. and Francis-Rose L.R. (2002). Progress in materials science. *The Martensitic Transformation in Ceramics—Its Role in Transformation Toughening*, 47(5), pp.463–557.

Kolitsidas C.I., Jonsson B.L., Persson P. and Stjerman A. (2014). Exploiting asymmetry in a capacitively loaded strongly coupled dipole array. In *2014 Loughborough Antennas and Propagation Conference (LAPC)* (pp.723–726), IEEE.C, Loughborough, https://doi.org/10.1109/LAPC.2014.6996497.

Kroics K., Sokolovs A., Sirmelis U. and Grigans L. (2016). Interleaved series input parallel output forward converter with simplified voltage balancing control. In *International Conference for Power Electronics, Intelligent Motion, Renewable Energy and Energy Management (PCIM Europe 2016)*, pp.10–12, IEEE Xplore.

Kuo W., Prasad V.R., Tillman F.A. and Hwang C.L. (2001). *Optimal reliability design*, Cambridge University Press, New York.

Leveson N. (2011). *Engineering a safer world: Systems thinking applied to safety*, MIT Press, Cambridge, MA.

Lewis E.E. (1996). *Introduction to reliability engineering*, Wiley, New York.

Li H., Wang J., Bai G. and Hu X. (2021). Research on self-balancing system of autonomous vehicles based on queuing theory. *Sensors MDPI*, 4619, pp.1–16.

Lindroos M., Apostol M., Heino V., Valtonen K., Laukkanen A., Holmberg K. and Kuokkala V.T. (2015). The deformation, strain hardening, and wear behaviour of chromium-alloyed hadfield steel in abrasive and impact conditions. *Tribology Letters*, 57(3), op24.

Mahamood R.M., Akinlabi E.T., Shukla M. and Pityana S. (2012). Functionally graded material: An overview. In *Proceedings of the World Congress on Engineering 2012, Vol III, WCE 2012, July* (pp.4–6), International Association of Engineers (IAENG), London.

Mämmelä O., Hiltunen J., Suomalainen J., Ahola K., Mannersalo P. and Vehkaperä J. (2016). Towards micro-segmentation in 5G network security. In *Proceedings of the EuCNC 2016 Network Management, QoS and Security Workshop* (pp.148–153), Athens.

Mandolini M., Pradel P. and Cicconi P. (2022). Design for additive manufacturing: Methods and tools. *Applied Sciences*, 12(13), 6548. https://doi.org/10.3390/app12136548.

Matthews C. (1998). *Case studies in engineering design*, Elsevier; Arnold, London.

McDermott R.E., Mikulak R.J. and Beauregard M.R. (2008). *The Basics of FMEA*, 2nd ed., CRC Press, Taylor and Francis, Boca Raton, FL.

Meeker W.Q. and Escobar L.A. (1998). *Statistical methods for reliability data*, Wiley, New York.

Melchers R.E. (2001). On the ALARP approach to risk management. *Reliability Engineering and System Safety*, 71(2), pp.201–208.

Meraz M.A., Yánez A., Jiménez C. and Pichardo R. (2005). Self-balancing system for rotating mechanisms. *Revista de la Facultad de Ingeniería – Universidad de Tarapacá*, 13(2), pp.59–64.

Meriam J.L, Kraige L.G. and Bolton J.N. (2016). *Engineering mechanics: Statics*, 8th ed., Wiley, Hoboken, NJ.

MIL-STD-1629A, (1977). *US department of defence procedure for performing a failure mode and effects analysis*, US Department of Defence, Washington, DC.

Mott R.L, Vavrek E.M. and Wang J. (2018). *Machine elements in mechanical design*, 6th ed., Pearson Education, New York.

Niku-Lari A. (1981). Shot-peening. In *The First International Conference on Shot Peening, Paris* (pp.1–27), 14–17 September, Pergamon Press, Oxford.

Norton R.L. (2006). *Machine design, an integrated approach*, 3rd ed., Pearson International, New York.

O'Connor P.D.T (2002). *Practical reliability engineering*, 4th ed., Wiley, New York.

Orloff M.A. (2006). *Inventive thinking through TRIZ: A practical guide*, 2nd ed., Springer, New York.

Orloff M.A. (2017). *ABC TRIZ Introduction to creative design thinking with modern TRIZ modelling*, Springer, Switzerland.

Pahl G., Beitz W., Feldhusen J. and Grote K.H. (2007). *Engineering design*, 3rd ed., Springer, Berlin.

Paris P.C. and Erdogan F. (1963). A critical analysis of crack propagation laws. *Journal of Basic Engineering*, 85, pp.528–534.

Paris P.C., Gomez M.P. and Anderson W.P. (1961). A rational analytic theory of fatigue. *The Trend in Engineering*, 13, pp.9–14.

Pecht M. (1996). Why the traditional reliability prediction models do not work – Is there an alternative. *Electronic Cooling*, 2(1), pp.10–12.

Pecht M., Dasgupta A., Barker D. and Leonard C.T. (1990). The reliability physics approach to failure prediction modelling. *Quality and Reliability Engineering International*, September/October, 6(4), pp.267–273.

Petroski H. (1994). *Design paradigms: Case histories of error and judgment in engineering*, Cambridge University Press, Cambridge.

Petrov V. (2020). *TRIZ basics: Theory of inventive problem solving*, Vladimir Petrov, Dunstable, UK.

Ramakumar R. (1993). *Engineering reliability, fundamentals and applications*, Prentice Hall, Englewood Cliffs, NJ.

Rantanen K. and Domb E. (2008). *Simplified TRIZ*, 2nd ed., Auerbach Publications, New York.

Reade C. (1989). *Elements of functional programming*, Addison-Wesley Longman, Reading, MA.

Rosenthal S., Maaß F., Kamaliev M., Hahn M., Gies S. and Tekkaya A.E. (2020). Lightweight in automotive components by forming technology. *Automotive Innovation*, 3, pp.195–209.

Salamatov Y. (1999). *TRIZ: The right solution at the right time*, Insytec B.V., Hattem.

Samuel A. and Weir J. (1999). *Introduction to engineering design: Modelling, synthesis and problem solving strategies*, Elsevier Butterworth-Heinemann, Oxford.

Savransky S.D. (2000). *Introduction to TRIZ methodology of inventive problem solving*, CRC Press, Boca Raton, MA.

Sedgewick R. (1992). *Algorithms in C++*, Addison-Wesley, Reading, MA.

Si S, Zhao J and Cai Z. (2020). Recent advances in system reliability optimiation driven by importance measures. *Frontiers of Engineering Management*, 7(3), pp.335–358.

Starovoytova D. (2021). TRIZ: *Everyone can learn to be creative*, Diana Starovoyotova, Dunstable, UK.

Su Q., Fu G., Yang J. and Li S. (2025). Experiment and numerical analysis of prestressed unequal-walled rectangular concrete-filled steel box beams. *Nature Portfolio, Scientific Reports*, 15(425). https://doi.org/10.1038/s41598-024-84473-2.

Sundararajan C. (1991). *Guide to reliability engineering: Data analysis, applications, implementations and management*, Van Nostrand Reinhold, New York.

Sunstein C.R. (2002). *Risk and reason: Safety, law, and the environment*, Cambridge University Press, Cambridge.

Svenson O. (1991). The accident evolution and barrier function (AEB) model applied to incident analysis in the processing industries. *Risk Analysis,* 11(3) pp.499–507.

Teall J.L. and Hasan I. (2002). *Quantitative methods for finance and investment*, Blackwell Publishing, Oxford.

Terninko J., Zusman A. and Zlotin B. (1998). *Systematic innovation: An introduction to TRIZ*, St. Lucie Press, BocaRaton, FL.

Thompson G. (1999). *Improving maintainability and reliability through design*, Professional Engineering Publishing, London.

Todinov M.T. (2000). Residual stresses at the surface of automotive suspension springs. *Journal of Materials Science*, 35, pp.3313–3320.

Todinov M.T. (2004). Reliability analysis and setting reliability requirements based on the cost of failure. *International Journal of Reliability, Quality and Safety Engineering*, 11(3), pp.273–299.

Todinov M.T. (2006a). Reliability analysis based on the losses from failures. *Risk Analysis*, 26(2), pp.311–335.

Todinov M.T. (2006b). Reliability analysis of complex systems based on the losses from failures. *International Journal of Reliability, Quality and Safety Engineering*, 13(2), pp.1–22.

Todinov M.T. (2007). *Risk-based reliability analysis and generic principles for risk reduction*, Elsevier, Amsterdam.

Todinov M.T. (2009). A comparative method for improving the reliability of components. *Nuclear Engineering and Design*, 239, pp.214–220.

Todinov M.T. (2013). New models for optimal reduction of technical risks. *Engineering Optimization*, 45(6), pp.719–743.

Todinov, M.T. (2014). Optimal allocation of limited resources among discrete risk-reduction options. *Artificial Intelligence Research*, 3(4), pp.15–27.

Todinov M.T. (2015). Reducing risk through segmentation, permutations, time and space exposure, inverse states and separation. *International Journal of Risk and Contingency Management*, 4(3), pp.1–21.

Todinov M.T. (2016). *Reliability and risk models: Setting reliability requirements*, 2nd ed., Wiley, Chichester.

Todinov M.T. (2017a). Improving reliability and reducing risk by separation. *International Journal of Risk and Contingency Management*, 6(4), pp.16–39.

Todinov M.T. (2017b). Reducing risk by segmentation. *International Journal of Risk and contingency Management*, 6(3), pp.27–46.

Todinov M.T. (2017c). Reducing risk through inversion and self-strengthening. *International Journal of Risk and contingency Management*, 6(1), pp.14–42.

Todinov M.T. (2018). Reliability and risk controlled by the simultaneous presence of random events on a time interval. *ASCE-ASME Journal of Risk and Uncertainty in Engineering Systems, Part B, Mechanical Engineering*, 4(2), 021003.

Todinov M.T. (2019a). *Methods for reliability improvement and risk reduction*. Wiley, Chichester.

Todinov M.T. (2019b). Domain-independent approach to risk reduction. *Journal of Risk Research*, 23(6), pp.796–810.

Todinov M.T. (2019c). Mechanisms for improving reliability and reducing risk by stochastic and deterministic separation. *Journal of Risk Research*, 22(4), pp.448–474.

Todinov M.T. (2019d). Reliability improvement and risk reduction by inequalities and segmentation. *Proceedings of the Institution of Mechanical Engineer Part O: Journal of Risk and Reliability*, 234(1), pp.63–73.

Todinov M.T. (2020). Reducing the risk of failure by deliberate weaknesses. *International Journal of Risk and Contingency Management*, 9(2), pp.33–53.

Todinov M.T. (2022a). Improving reliability by increasing the level of balancing and by substitution. *Proceedings of the Institution of Mechanical Engineer Part C: Journal of Mechanical Engineering Science*, 237(7), pp. 1550–1562.

Todinov M.T. (2022b). A general class of algebraic inequalities for generating new knowledge and optimising the design of systems and processes. *Research in Engineering Design*, 33, pp.161–171.

Todinov M.T. (2024a). Lightweight Designs and Improving the Load-Bearing Capacity of Structures by the Method of Aggregation, *Mathematics*, 12, p.1522.

Todinov M.T. (2024b). Reverse engineering of algebraic inequalities for system reliability predictions and enhancing processes in engineering. *IEEE Transactions on reliability*, 73(2), pp.902–911.

Todinov M.T. (2025a). *Reverse engineering of algebraic inequalities, Exploring new physical realities and optimizing products and processes*, 2nd ed., CRC Press, Florida.

Todinov M.T. (2025b). Improving the reliability of parallel and series–parallel systems by reverse engineering of algebraic inequalities. *Mathematics*, 13, p.1381.

Todinov M.T. (2025c). Improving the load-carrying capacity of structures consisting of multiple elements. *Sustainable Structures*, 5(3), 000079.

Todinov M.T. (2025d). A new domain-independent method for improving reliability and reducing risk based on exploiting asymmetry. In *Proceedings of the Institution of Mechanical Engineers Part O: Journal of Risk and Reliability*. https://journals.sagepub.com/doi/full/10.1177/1748006X251324081.

Trivedi K.S. (2002). *Probability and statistics with reliability, queuing and computer science applications*, 2nd ed., Wiley, Chichester.

Ugural A.C. (2022). *Mechanical engineering design* (SI Edition), CRC Press, Boca Raton, FL.

Uicker J.J. Jr., Pennock G.R. and Shigley J.E. (2003). *Theory of machines and mechanisms*, 3rd ed., Oxford University Press, New York.

US patent 3814962A (1974). *Magnetic worm drive,* inventor: M. Baermann, United States Patent Office, US.

US patent 4072159 (1978). *Emergency valve incorporating thermal foamable plastic material*, inventor: Toyoki Kurosava, United States Patent Office, US.

US patent 3528206A (1968). *Thermal expansion compensation device*, inventor: Kenneth M. Baird, United States Patent Office, US.

US patent US2011/0027007A1 (2011). *Retrofittable cable mechanical fuse*, inventor: David Tunno and Stuart Larsen, United States Patent Office, US.

US patent US3693698A (1969). *Method of casting volatile metals*, inventors: A. Balevski and I. Dimov, United States Patent Office, US.

Vose D. (2000). *Risk analysis, a quantitative guide*, 2nd ed., Wiley, New York.

Walton D. and Moztarzadeh H. (2017). Design and development of an additive manufactured component by topology optimization. Procedia CIRP 60, 205–210. *Chinese Journal of Aeronautics*, 33(12), pp.3206–3219. https://doi.org/10.1016/j.procir.2017.03.027.

Wang S., Bojnordi M.N., Guo X. and Ipek E. (2018). Content aware refresh: Exploiting the asymmetry of DRAM retention errors to reduce the refresh frequency of less vulnerable data. *IEEE Transactions on Computers*, 68(3), pp.362–374.

Wirth N. (1976). *Algorithms+data structures = programs*, Prentice-Hall, Englewood Cliffs, NJ.

Yang L., Harrysson O.L.A., Cormier D., West H., Zhang S., Gong H. and Stucker B. (2016). Design for additively manufactured lightweight structure: A perspective, solid freeform fabrication 2016. In *Proceedings of the 26th Annual International Solid Freeform Fabrication Symposium – An Additive Manufacturing Conference.*

Yi Y.D., Hu Y. and Li D. (2021). Redundancy optimization for multi-performance multi-state series-parallel systems considering reliability requirements. *Reliability Engineering and System Safety*, 215, 107873.

Zhang L., Konyukhov A., Mok E. and Choo H. (2013). Increasing the road safety of e-bike: design of protecting shell based on stability criteria during severe road accidents. In *Proceedings of the 2013 International conference on electrical, control and automation engineering (ECAE '13)* (pp.485–489).

Zhao Z., Kang Z., Zhang T., Zhao B., Zhang D. and Yan R. (2024). Topology optimization algorithm for spatial truss based on numerical inverse hanging method. *Journal of Constructional Steel Research*, 219, 108764.

Zhen J., Ding L. and Wang X. (2017). Design of the smart pneumatic balance crane control system. In *5th International Conference on Mechatronics, Materials, Chemistry and Computer Engineering (ICMMCCE 2017*).

Zhou Y., Uy B., Wang J. and Dongxu L.D. (2021). Behavior and design of welded stainless steel beams with compact sections under flexure and shear. *Journal of Constructional Steel Research*, 187, 106996.

Zhu J., Zhou H., Wang C., Zhou L., Yuan S. and Zhang W. (2021). A review of topology optimization for additive manufacturing: Status and challenges. *Chinese Journal of Aeronautics*, 34(1), pp.91–110.

Index

For Product Safety Concerns and Information please contact our EU
representative GPSR@taylorandfrancis.com
Taylor & Francis Verlag GmbH, Kaufingerstraße 24, 80331 München, Germany

www.ingramcontent.com/pod-product-compliance
Ingram Content Group UK Ltd.
Pitfield, Milton Keynes, MK11 3LW, UK
UKHW021514250926
475702UK00043B/493
* 9 7 8 1 0 4 1 1 3 9 3 4 8 *